Electronic Materials

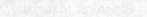

Electronic Materials

L. A. A. Warnes
Department of Electronic and Electrical Engineering
Loughborough University of Technology

VNR VAN NOSTRAND REINHOLD
_____New York

Copyright © 1990 by L. A. A. Warnes

ISBN 0–442–30477–3

Published in the United States of America by
Van Nostrand Reinhold
115 Fifth Avenue
New York, New York 10003

Distributed in Canada by
Nelson Canada
1120 Birchmount Road
Scarborough, Ontario M1K 5G4, Canada

Published in Great Britain by
Macmillan Education Ltd
Houndmills, Basingstoke, Hampshire RG21 2XS
and London

Printed in Singapore

16 15 14 13 12 11 10 9 8 7 6 5 4 3 2 1

Library of Congress Cataloging-in-Publication Data
Warnes, L.A.A.
   Electronic materials/L.A.A. Warnes.
     p.   cm.
   Includes index
   ISBN 0–442–30477–3
   1. Electronics—Materials.   I. Title
TK7871.W34   1990
621.381—dc20

                              90–40593
                                  CIP

Contents

Preface

The importance of materials science for the progress of electronic technology has been apparent to all since the invention of the transistor in 1948, though that epoch-making event was the result of far-sighted research planning by Bell Laboratories dating from a decade or more before: no mere chance discovery, therefore, but the fruition of work which allotted at its inception a vital rôle to materials. The transistor is now very old hat, but new materials developments are continually triggering fresh developments in electronics, from optical communications to high-temperature superconductors.

Electronic engineers are now given at least two courses in materials as part of their degree programme. This book arose from a series of forty lectures the author gave to the third year students on the Extended Honours Degree Course in Electronic and Electrical Engineering at Loughborough University, though additional elementary material has been included to make the book suitable for first year students. The biggest problem in such a course is deciding what must be left out, and this I am afraid I shirked by leaving out all those areas which I was not familiar with from my days in the Ministry of Aviation, the semiconductor device industry and as a graduate student and research worker. I hope that what remains is sufficiently catholic. Where possible I have tried to use the problems at the end of each chapter to expand the text, but stock questions are needed also. Quite a number of the problems are those which arose in practical situations: they are often more interesting because of the manner of their genesis, but also more difficult.

To my wife, Irene, I offer my thanks for putting up with this for longer than I like to recall.

Frequently-used Physical Constants

c, the speed of light in a vacuum	3.00×10^8 m/s
$-e$, the electronic charge	-1.60×10^{-19} C
g, the acceleration due to gravity	9.81 m/s^2
h, Planck's constant	6.63×10^{-34} Js
$\hbar$, Planck's constant divided by 2π	1.055×10^{-34} Js
k_B, Boltzmann's constant	1.38×10^{-23} J/K
m_0, the electronic rest mass	9.11×10^{-31} kg
at., the standard atmosphere	1.013×10^5 Pa
a.m.u., the atomic mass unit	1.66×10^{-27} kg
ϵ_0, the permittivity of a vacuum	8.85×10^{-12} F/m
μ_0, the magnetic constant	$4\pi \times 10^{-7}$ H/m
μ_B, the Bohr magneton	9.27×10^{-24} J/T

1

The Structure of Solids

1.1 Ideal Crystal Structures

Many important properties of solids depend on their crystal structures. Fortunately, relatively few of these need concern us: about three-quarters of the elements, for instance, crystallize in one or more of three simple structures, namely face-centred cubic (fcc, or cubic close-packed), body-centred cubic (bcc) or hexagonal close-packed (hcp). Two of these were suggested by W. Barlow in a remarkable paper of 1883, 30 years before W. L. Bragg published the first crystal structure to be determined by X-rays.

1.1.1 The Close-packing of Spheres

Barlow said that if there were hard, spherical atoms in a solid, they could be arranged in layers of closest packing as in figure 1.1(a). The symmetry of this is hexagonal, more clearly seen in figure 1.1(b), where the centres of the atoms have been replaced by points. A three-dimensional structure can then be assembled by stacking identical layers on top of one another. There are three ways of doing this, leading to three different structures. If the first layer is denoted A, then the next layers can be placed immediately above in an identical orientation to form the stacking sequence AAA... , a simple hexagonal structure. However, the structure is not one of closest packing, since it is possible to arrange the layer above A to fill either the hollows marked B or those marked C, producing a greater packing density for the atoms (see problem 1.1). Supposing the second layer occupies the B-sites, then the next layer above can take up the same orientation as the first to form the stacking sequence ABABA... . This structure is called hexagonal close-packed or hcp for short. Alternatively, the third layer can take up position C before the sequence of layers is repeated, to form the stacking sequence ABCABCA... . Surprisingly, this structure has cubic symmetry and is known as fcc or cubic close-packed. In both hcp and fcc

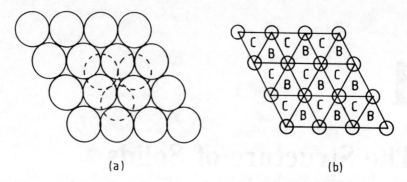

(a)   (b)

Figure 1.1   *(a) A layer of close-packed spheres; (b) conventional representation of close-packed spheres*

metals there are 12 nearest neighbours for each atom: the atomic environments are identical. The description of crystal structures is made much easier with the help of a few concepts, which we shall now consider.

1.1.2 The Lattice and the Basis

A *lattice* is an array of points in space related by symmetry. For cubic close-packed spheres, the lattice is obtained in the simplest possible way – each sphere is merely replaced by a point at its centre. Figure 1.2 shows the

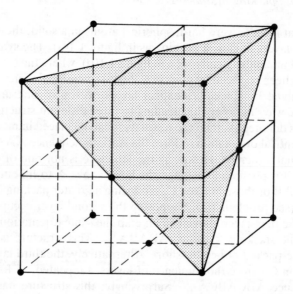

Figure 1.2   *The fcc structure*

resulting fcc lattice, in which lines have been drawn between the lattice points to make the symmetry clearer: they form no part of the lattice. The shaded plane is one of the close-packed planes shown in figure 1.1(b). Were we to replace each point of the fcc lattice with a nickel atom centred on that point, then the crystal structure of nickel would be the consequence. The *basis* of the structure is said to be one nickel atom. Put simply,

LATTICE + BASIS = STRUCTURE

There is no need to place the centre of the atoms at the lattice point: so long as each atom is placed in precisely the same relationship to each lattice point, the same structure results. Usually the basis comprises more than one atom, even in elemental structures such as silicon, while compounds *must* have a polyatomic basis; for example, sodium chloride has a basis of one NaCl molecule with an fcc lattice. A Cl^- ion is at each lattice point with an Na^+ ion halfway between, producing the structure known variously as the halite, rocksalt or NaCl structure (figure 1.3):

fcc lattice + NaCl molecule (basis) = halite structure

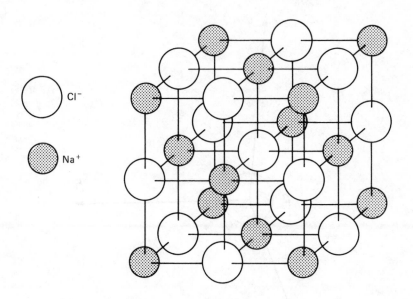

Figure 1.3 *The NaCl structure*

1.1.3 The Unit Cell

A crystal of any given size may be formed by the repetition in identical orientation of a part of the structure. The smallest convenient part that may be so employed is the *unit cell*. Within the unit cell, directions are related to *a*-, *b*- and *c*-axes, and the dimensions of the unit cell along these axes are *a*, *b*, and *c*, which are the repeat distances of the structure, or *lattice parameters*. Lattice parameters are usually given in Ångstroms (1 Å = 0.1 nm) in the literature; we shall do likewise for convenience. The angles between the axes are α, β and γ as in figure 1.4. For a cubic unit cell, $a = b = c$ and $\alpha = \beta = \gamma = 90°$, so the cell volume is just a^3. A corner of the unit cell is chosen as the origin and within the cell positions are given as fractions of the lattice parameters, so an atom in the centre of the cell at $\frac{1}{2}a$, $\frac{1}{2}b$, $\frac{1}{2}c$, is said to be at $\frac{1}{2}, \frac{1}{2}, \frac{1}{2}$. The order is invariable: *a*-, *b*-, *c*-distance.

Few can draw and understand three-dimensional crystal structures, so we shall reduce them to two-dimensional plan views, looking down the *c*-axis. Atoms in the unit cell are projected onto the plane of the paper, with the *c*-axis perpendicular to it. For example, the fcc unit cell of aluminium is drawn in figure 1.5. To show which atoms are not in the plane of the paper, they are either drawn part shaded (half for an atom $\frac{1}{2}c$ up from the plane of the paper and so on) or a fraction is written next to them,

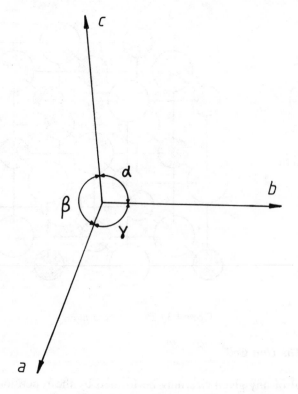

Figure 1.4   *The crystallographic axes*

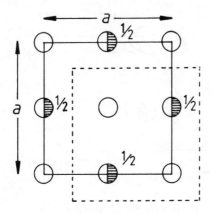

Figure 1.5 *The fcc unit cell of Al in two dimensions*

expressing the same thing. From the plan view of the unit cell the number of atoms in the cell is found by drawing the outline of the cell, moving it diagonally as shown by the dotted lines and counting the atoms inside the dotted outline. There are obviously four atoms in the unit cell of aluminium. The density of aluminium can then be calculated from the lattice parameter and relative atomic mass (r.a.m., given in the Periodic Table at the end of the book). The unit cell volume is a^3, so

$$\rho = nW \times 1.66 \times 10^{-27}/a^3$$

where n is the number of atoms/unit cell, W is the r.a.m. and one atomic mass unit weighs 1.66×10^{-27} kg. Substituting for W and a (4.05 Å) leads to $\rho = 2700$ kg/m^3

In the case of hexagonal crystals the unit cell axes are not all perpendicular; the a- and b-axes make an angle of 120° with each other and the c-axis is perpendicular to these. Figure 1.6 shows the unit cell of magnesium in plan view. The lattice is hexagonal, but the unit cell is a parallelepiped which forms one-third of a hexagonal prism. There is a lattice point at each corner of the unit cell and associated with each lattice point are two magnesium atoms – one at the lattice point and one at $\frac{2}{3}, \frac{1}{3}, \frac{1}{2}$. The number of atoms in the unit cell is still found the same way; there are just two this time. However, the unit cell volume is now ahc, where h is the height of the parallelogram in figure 1.6, which must be $\frac{1}{2}a\sqrt{3}$. The axial ratio (c/a) is, ideally, $\sqrt{(8/3)}$, but few hcp metals do have the ideal axial ratio.

A very common structure for metals is *body-centred cubic* (the alkali metals lithium, sodium etc. crystallize in this), whose unit cell is shown in figure 1.7(a). It has lattice points at the corners and at the centre of the unit cell, while the basis is an atom at each lattice point. There are two atoms to the unit cell, which is not a close-packed structure (see problem 1.4). The

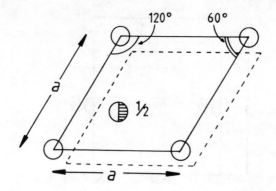

Figure 1.6   *The hcp unit cell of Mg*

simple cubic unit cell shown in figure 1.7(b) has just one lattice point at each corner. If we add to this basis a CsCl molecule, so that the Cl^- ions are at the lattice points and the Cs^+ ions are at $\frac{1}{2}, \frac{1}{2}, \frac{1}{2}$ (that is, the cell centre), we get the caesium chloride structure of figure 1.7(b), which is NOT to be confused with bcc, since the ions are not identical.

1.1.4   Miller Indices

Planes and directions in crystals may be succinctly denoted by sets of numbers called *Miller indices*. Consider the plane shown in figure 1.8. It

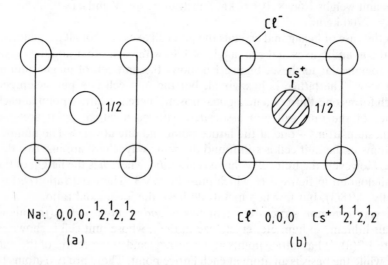

Figure 1.7   *(a) The bcc unit cell of Na; (b) the simple cubic unit cell of CsCl*

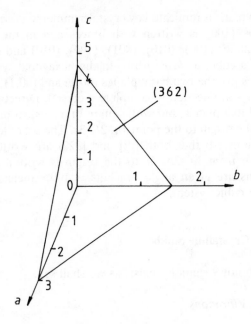

Figure 1.8   *The (362) plane*

makes intercepts on all three axes – at $3a$ units on the a-axis, $1\frac{1}{2}b$ units on the b-axis and $4\frac{1}{2}c$ on the c-axis. The Miller indices of the plane are found by first taking reciprocals of the axial intercepts (ignoring the lattice repeat units a, b and c): $1/3$, $1/1\frac{1}{2}$, $1/4\frac{1}{2}$. These numbers are all then multiplied by the smallest number which will turn them into integers, in this case nine, to form the triad (362), which are the Miller indices of the plane. The notation must be used precisely thus: parentheses, and the invariable order, a-, b-, c-axis reciprocal intercept. Most planes of interest have indices like 0, 1 or 2 and are known as *low index* planes. Negative intercepts are written with a bar over them thus: $(\bar{1}2\bar{3})$. If a plane is parallel to a given axis, then its intercept is at infinity and its reciprocal intercept is therefore zero; so a zero in the Miller indices of a plane means that it must be parallel to that axis. For example, the (001) plane is parallel to a- and b-axes and intercepts the c-axis.

   In the hexagonal crystal system it is customary to use three axes in the basal plane (the plane of the paper when we draw the unit cell), with 120° between them. This gives rise to four Miller indices for a plane, the first three for the basal plane axes and the fourth for the c-axis. However, the third index is related to the first two by geometry, so that in the case of the $(hkil)$ plane $i = -(h + k)$. This dummy index is often omitted and replaced by a dot thus: $(hk.l)$, or (11.0) for (11$\bar{2}$0).

When a given plane is replicated by crystal symmetry a *form* is the result. A form such as {100} is written with braces and in the cubic system comprises the planes (100), (010), (001), ($\bar{1}$00), (0$\bar{1}$0) and (00$\bar{1}$), making the six faces of a cube, a *closed* form. In the hexagonal system the form {00.1} comprises just the two basal planes (00.1) and (00.$\bar{1}$), an *open* form since other planes are needed to complete a crystal. Directions in crystals are represented like planes, but with square brackets, so that [123] is the direction from the origin to the point 1a, 2b, 3c. The set of least integers is used for directions, so that both [$\frac{111}{222}$] and [222] are written [111]. The reverse direction from 1a, 2b, 3c to the origin is written [$\bar{1}\bar{2}\bar{3}$]. Sets of crystal directions are given angle brackets: <100> means [100], [010], [001] etc. in the cubic system.

1.2   Defects in Crystalline Solids

Ideal crystals do not – cannot – exist, as we shall see.

1.2.1   Thermal Vibrations

The temperature of a solid is a measure of the amplitude of the random vibration of its atoms and these vibrations cause the displacement of the atoms from their ideal positions. This displacement is an important cause of electrical resistance in metals: without it pure metals would have very much less resistance than they do have. At room temperature, for example, copper has about a hundred times the resistivity it has at liquid helium temperature (4.2 K), where thermal vibrations have almost ceased. Other crystal defects may be classified by their dimensionality.

1.2.2   Zero-dimensional (Point) Defects

There are three kinds of point defect in crystalline solids.

(1) Vacancies, the sites where atoms are missing.
(2) Substitutional impurities, which are foreign atoms that replace host atoms.
(3) Interstitial atoms, or interstitialcies, are atoms located in the holes between atoms of the host lattice. They may be foreign atoms or host atoms that have become misplaced.

Vacancies

When a crystal forms, usually at high temperatures, many vacancies are present. As the crystal cools, the equilibrium number of vacancies falls too, by a process of diffusion. Eventually the rate of diffusion becomes so slow that the vacancies are frozen into the structure. Thus at room temperature, the vacancies in a solid are not at equilibrium, but are in concentrations characteristic of some higher temperature. Vacancies are important because they are responsible for diffusion and void reduction in solids (see section 1.2.5). They play a minor role in the electrical resistivity of pure metals.

Suppose the energy required to form a vacancy is E_v. Then the equilibrium number of vacancies, n, at an absolute temperature, T, is given by the Boltzmann distribution

$$n = N\exp(- E_v/k_B T) \qquad (1.1)$$

where N is a constant and k_B is Boltzmann's constant. A plot of $\ln(n)$ against $1/T$ will give a straight line of slope $- E_v/k_B$, sometimes known as an Arrhenius plot. Now n can be measured by indirect means fairly simply: for example, by measuring the thermal expansion ($\delta l/l$) of a sample as the temperature rises and comparing it with the change in lattice parameter found by X-rays ($\delta a/a$). The X-rays do not take any notice of the odd vacant site, so the change in volume that they record is smaller than the actual change in volume. Alternatively, the sample can be quenched (cooled rapidly) from high temperature to freeze in the vacancies, and the density at room temperature of the quenched sample can be compared to that of a slowly cooled sample.

Figure 1.9 shows data for aluminium taken from *Phys. Rev.* **117**, 52 (1960), in which the $\ln(\delta l/l - \delta a/a)$ is plotted against $1/T$. The vacancy concentration will be proportional to ($\delta l/l - \delta a/a$), since $\delta V/V = 3\delta l/l$. The slope of the line in figure 1.9 is $- 9700$, so

$$E_v/k_B = 9700$$
$$E_v = 9700 \times 1.38 \times 10^{-23} \text{ J} = 0.84 \text{ eV}$$

1 eV is 1.6×10^{-19} J, the electronic charge (1.6×10^{-19} C) $\times$ 1 V. For convenience, energies in atoms are given in eV.

The vacancy concentration at the melting point can be estimated from figure 1.9 to be about 0.1 per cent, a value typical of most metals at their melting points. Vacancies give rise to an anomalous increase in specific heat, which may also be used to find their concentrations and energies of formation (see problem 1.5).

Substitutional and Interstitial Atoms

When an impurity atom is dissolved in a solid it can take up a normal lattice site by displacing a host atom, or it can squeeze into the spaces between host atoms: which it does depends largely on the relative sizes of host and impurity atoms. In the case of ionic solids the position is complicated by

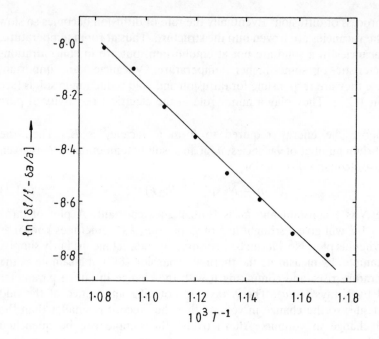

Figure 1.9 *The anomalous thermal expansion of Al*

the requirement for charge neutrality. If $CaCl_2$ is dissolved in KCl, for example, the Ca^{2+} ions replace the K^+ ions, but because there must be charge neutrality in the solid, a vacant K^+ site must be formed at the same time. In Cu_2O, there is a small fraction of Cu^{2+} ions each requiring a Cu^+ ion vacancy to maintain charge neutrality. The extra positive charge on the cations is mobile and is responsible for p-type conductivity in the material: Cu_2O is termed a defect semiconductor.

It can be shown that the work done in expanding a spherical void in a continuum is given by

$$W = 6GV_0\epsilon^2 \qquad (1.2)$$

where G is the shear modulus, V_0 is the original volume of the void and ϵ is the strain ($\delta l/l$). Suppose an atom has a radius 10 per cent larger than copper, for which $G = 50$ GPa. Looking at figure 1.10, which shows part of the unit cell of copper, we can see that the copper atoms touch along the face diagonals, so that $4r = a\sqrt{2}$, where r is the radius of a copper atom and $a\sqrt{2}$ is the length of the unit cell diagonal. Thus $r = \frac{1}{4}a\sqrt{2} = 0.3535 \times 0.362 = 1.28$ Å, and the volume of a copper atom is $\frac{4}{3}\pi r^3$, or 8.78×10^{-30} m³. Calling this the initial volume, V_0 in (1.2) and setting $\epsilon = 0.1$ (10 per cent) leads to

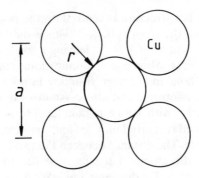

Figure 1.10   *The unit cell of Cu*

$$W = 6 \times 50 \times 10^9 \times 8.78 \times 10^{-30} \times (0.1)^2$$
$$= 2.6 \times 10^{-20} \text{ J} = 0.16 \text{ eV}$$

This is a fairly small energy, so we can expect a good deal of solid solubility. Aluminium atoms are about 10 per cent larger than those of copper and its solid solubility in copper is about 20 atomic per cent.

The same sort of strain energy consideration applies to interstitial atoms. For instance, suppose we want to dissolve carbon in iron. At room temperature iron is bcc with lattice parameter 2.87 Å and the touching atoms are at $0, 0, 0$ and $\frac{1}{2}, \frac{1}{2}, \frac{1}{2}$, that is, at the cell corners and the cell centre (see figure 1.11). Thus the distance between atoms is $\sqrt{(\frac{1}{4}a^2 + \frac{1}{4}a^2 + \frac{1}{4}a^2)} = \frac{1}{2}a\sqrt{3} = 2.485$ Å, which is defined as the diameter of an iron atom. The

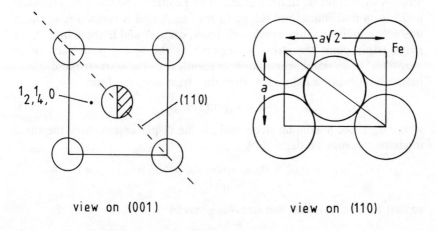

view on (001)   view on (110)

Figure 1.11   *A view of the bcc Fe unit cell on (110)*

largest space in the bcc structure is centred on the point $\frac{1}{2}, \frac{1}{4}, 0$ – which is $\sqrt{[(\frac{1}{2}a)^2 + (\frac{1}{4}a)^2]}$ from the atom at $0, 0, 0$. This is only 1.6 Å, and the iron atom at $0, 0, 0$ has a radius of 1.24 Å, leaving a space only $2(1.6 - 1.24)$, or 0.72 Å, in diameter. Carbon has an atomic radius of about 0.7 Å, so it is twice as big as the 'hole' in bcc iron, resulting in poor solubility.

Iron, however, is allotropic and also crystallizes in the fcc structure at elevated temperatures, with lattice parameter 3.65 Å. The largest interstice in the fcc unit cell is centred on $\frac{1}{2}, \frac{1}{2}, \frac{1}{2}$ and the nearest atoms are at the face centres, $\frac{1}{2}, 0, \frac{1}{2}$ etc. The distance between the two is therefore $\frac{1}{2}a$. As we have already seen, the atoms in the fcc unit cell touch along the face diagonals, so they are $\frac{1}{2}a\sqrt{2}$ in diameter and $\frac{1}{4}a\sqrt{2}$ in radius, leaving a space of $2(\frac{1}{2}a\sqrt{2} - \frac{1}{4}a\sqrt{2})$, or 1.07 Å, for any interstitial atom to fit into. Though this is still smaller than the diameter of the carbon atom (1.4 Å), it is much less of a squeeze and carbon is quite soluble in fcc iron. Steelmaking largely depends on this fact.

Because the strain energy goes as the square of the strain, and the concentration of interstitial or substitutional defects goes as $\exp(-W/k_B T)$, it is clear that the concentration is very sensitive to the mismatch between the host lattice and the impurity. Experimentally, the energy of carbon interstitialcies is 0.3 eV in fcc iron and 0.8 eV in bcc iron (see problem 1.13).

1.2.3 *One-dimensional (Line) Defects: Dislocations*

In 1926 Frenkel made a simple, but reasonable, estimate of the critical shear stress of a solid, which is the stress required to move two planes of atoms over each other. Figure 1.12 shows two such planes from sideways on and the stress–displacement graph corresponding to the movement of atom A over atom B. In their equilibrium positions the stress is zero, then builds to a maximum until falling to zero as A and B reach a position of unstable equilibrium and maximum strain, when A and B are in line. After passing this point the stress is negative as the atoms push each other 'downhill' to an equilibrium position identical to the start. If the distance between atoms in a plane is d, then the stress is of the form

$$\sigma = \sigma_m \sin(2\pi\xi/d) \tag{1.3}$$

where σ_m is the maximum stress and ξ is the displacement. Now the shear modulus, G, may be defined by

$$G = \text{shear stress/shear strain}$$
$$= \sigma/(\xi/y)$$

so that for small strains, the stress is given by

$$\sigma = G\xi/y \tag{1.4}$$

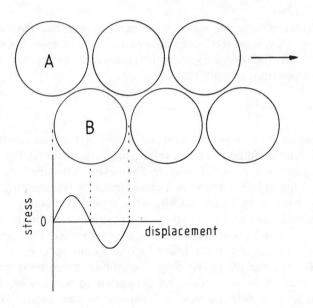

Figure 1.12 *Shear stress for two planes of atoms sliding over one another*

where y is the distance between planes. When ξ is small (1.3) becomes

$$\sigma = 2\pi\sigma_m\xi/d$$

Comparing this with (1.4), $\sigma_m = Gd/2\pi y$. Now $d = y$, approximately, so σ_m should be about $G/6$. If a stress this size is applied to a material, it will yield by the slippage of one plane over the other until it breaks.

The critical stress should be very large: G for a typical metal is about 50 GPa, so σ_m should be about 8 GPa. In fact the point at which metals begin to yield by plastic deformation is very much lower than this and is very dependent on the microstructure and composition of the sample under test. Pure metals yield at lower stresses than alloys, in general, and single crystals yield at lower stresses than polycrystalline samples of the same material. The very highest values of yield stress in metals are about $G/100$ and the lowest $G/10\,000$. To be sure, these facts were known to Frenkel among others and endeavours were made to reconcile theory and experiment by taking into account the structure of the metal. Still, the best that could be done was to lower the critical stress to about $G/30$, which is somewhere near the very highest that could be obtained in practice. Matters remained in this embarrassing state until 1934, when Taylor, Orowan and Polanyi put forward independently the idea that dislocations might be responsible for the low yield stresses observed in metals. The misconception in the early yield–strength calculations lay in assuming that crystal structures undergoing strain were perfect, whereas Taylor *et al.* suggested that they were not, but contained a type of one-dimensional defect known as the dislocation, which caused a misregistry of the atom-to-atom positions along a line.

Types of Dislocation

Figure 1.13(a) shows a cross-section through part of a crystal in which an extra plane of atoms (indicated by an arrow) has been inserted. The end of the inserted plane is the line of an *edge dislocation*. The structure of the crystal around the dislocation is shown more clearly by representing planes of atoms by lines as in figure 1.13(b). If a crystal containing an edge dislocation is subjected to a shear stress, the movement of the dislocation through the crystal will eventually cause the complete slip of the upper part of the crystal with respect to the lower by one atomic spacing, as shown in figure 1.14. For the plastic yielding to continue, there must be more dislocations present or they must be generated in some way. Metals are very adept at generating new dislocations so that slip is virtually continuous.

There is another type of dislocation (shown in figure 1.15), the *screw dislocation*. This is most easily envisaged by assuming that the crystal in the diagram has been cut and then sheared along the line AB, the material either side of the cut being joined afterwards. The crystal structure a few atoms' width from AB is essentially perfect and unstrained, while maximal strain is experienced along AB, which is the line of a screw dislocation. Dislocations usually are not purely edge or screw but a mixture of both.

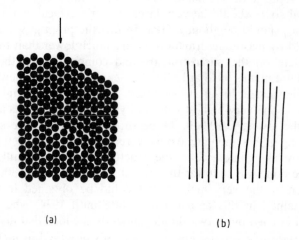

(a)   (b)

Figure 1.13   *(a) An edge dislocation; (b) a representation of an edge dislocation using line for planes of atoms*

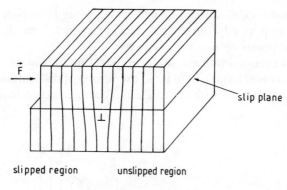

slipped region   unslipped region

symbol for edge dislocation: ⊥

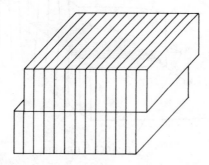

Figure 1.14   *How an edge dislocation produces shear*

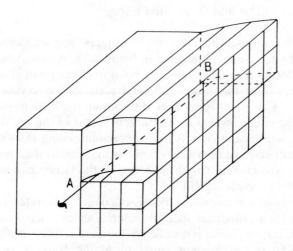

Figure 1.15   *A screw dislocation*

Figure 1.16 shows a dislocation line in a crystal going from pure screw on the front face to pure edge on the side face, and which is partly edge and partly screw at points between.

Dislocations cannot terminate inside a crystal, but only at its surface; or they may form closed loops within it. So the density of dislocations is stated as the total length per unit volume or the number intersecting unit area of a given plane.

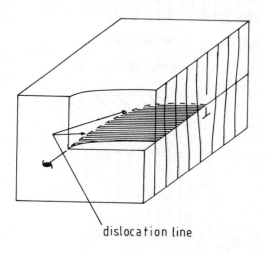

dislocation line

Figure 1.16   *A dislocation partly edge dislocation and partly shear dislocation*

The Burgers Vector and Dislocation Energy

Dislocations are identified by their Burgers vectors, which may be explained by the device illustrated in figure 1.17. A closed path is traced about a dislocation going from lattice point to lattice point in an anticlockwise sense, while the same path is traced in the crystal without a dislocation. The path about the dislocation requires a vector, b, to complete the circuit, which is the Burgers vector of the enclosed dislocation. A screw dislocation has its Burgers vector pointing along its line while edge dislocations have their Burgers vectors normal to their lines. Burgers vectors are written as a length, in terms of the lattice parameter, and a direction, for example $\frac{1}{2}a[110]$.

The energy of a screw dislocation is calculated with reference to figure 1.18, in which a cylindrical shell of material about a screw dislocation is considered. If the cylinder is unrolled as in figure 1.18(b), it is seen to have been sheared by an amount equal to b, the Burgers vector of the dislocation. Now the energy stored per unit volume (U_v) in a material

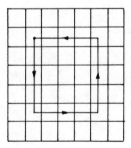 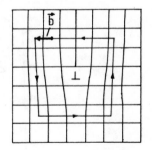

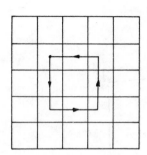 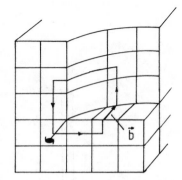

Figure 1.17   *The Burgers vector*

which has undergone strain is $\frac{1}{2}\sigma\epsilon$ where σ is the stress and ϵ is the strain. In this case $\sigma = G\epsilon$, so that $U_v = \frac{1}{2}G\epsilon^2$, where the shear strain, ϵ, is $b/2\pi r$, as can be seen from figure 1.18(b). Thus

$$U_v = \delta U/\delta V = \tfrac{1}{2}G(b/2\pi r)^2$$

The volume, δV, of the cylindrical shell is $2\pi r l\delta r$ and then

$$\delta U = U_v\delta V = \tfrac{1}{2}G(b/2\pi r)^2 \times 2\pi r l\delta r$$

so

$$U = \int dU = \int_{R_0}^{R} Gb^2 l/(4\pi r)dr$$

The upper limit, R, is the perpendicular distance from the dislocation line to where there is no appreciable strain (or the boundary of the crystal) and the lower limit, R_0 is about one atom's radius. Integrating

$$U = (Glb^2/4\pi)\ln(R/R_0)$$

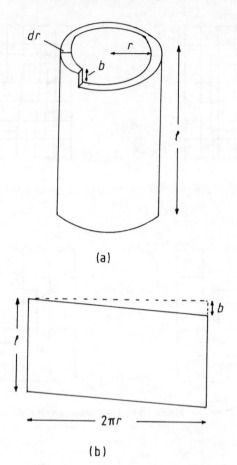

(a)

(b)

Figure 1.18   *(a) Shear of a cylindrical shell around a screw dislocation; (b) the cylinder*
in (a) unwound

The energy per unit length of dislocation is then

$$U_1 = Gb^2 \qquad (1.5)$$

where $\ln(R/R_0)$ has been taken to be 4π, as it gives the right order of
magnitude for U_1, and is a convenient number. In a typical metal
$G = 50$ GPa and $b = 2$ Å, making $U_1 \approx 2$ nJ/m.

Finding the energy of an edge dislocation is straightforward (see
problem 1.7), and it is about the same as that of a screw dislocation. If the
dislocation is 1 mm long, a reasonable value, then its total energy is about
2 pJ, or 12 MeV. Such a large energy means that the equilibrium density of
dislocations would be zero at any temperature up to the melting point.
Why then are dislocations present? (Even in the most carefully prepared

crystals, such as device-grade silicon, there are about $10^5/m^2$.) One reason is that crystals cannot grow on a perfect surface and the dislocations which nucleate growth are propagated in the growing crystal; another reason is that they are readily generated at high temperatures by thermal stresses, which are inevitable when a crystal is formed. These dislocations are then frozen in place when the crystal is cooled, ready to be set in motion by a suitable applied stress. Dislocation densities can be as high as $10^{16}/m^2$ in heavily cold-worked metals.

The Motion of Dislocations and the Strength of Materials

The motion of dislocations is responsible for the low plastic yield strengths of metals, as we have seen; paradoxically, however, dislocation motion is the reason metals are so useful to us, because without them metals would be brittle and difficult to form or use. In non-metals dislocations are less readily accommodated by the crystal lattice because the nature of the bonding is directional. The dislocations are immobile at normal temperatures, but may be activated if the temperature is raised enough. Most solids therefore have a ductile–brittle transition temperature, such as 100 K for mild steel or 1500 K for alumina. A material under tensile stress usually fails by the growth of a crack. In brittle materials, the material at the crack tip cannot deform plastically to relieve the stress and failure is not only catastrophic, but also occurs at an unpredictable stress. In a metal the crack tip raises the stress merely to the point where local plastic deformation can occur to relieve it. Failure is less sudden and more predictable.

Having seen that the motion of dislocations leads to low yield stresses, it is easy to see that impeding that motion will raise them: the question is – how? One way, we have all observed – cold work. If a piece of new copper pipe is bent sharply a few times, it soon becomes much more resistant to deformation, but heating it in a gas flame for a moment or two restores its ductility. What has happened is that during cold work dislocations have multiplied enormously so that they run into each other and prevent movement. Annealing at a high temperature removes the kinks and jogs of intersecting dislocations which prevent their movement and softens the metal. Impurities are another method of stopping dislocations from moving. These work in two ways: firstly, by forming fine precipitates which pin dislocations and secondly, by remaining in solid solution and binding themselves to dislocations so that much energy is required to move the dislocation away from the impurities. Because foreign atoms are better accommodated at dislocations they provide a simple method of detection by 'decorating', in which impurity atoms are dissolved in a crystal and accumulate at the sites of dislocations, thus rendering them visible. Dispersion strengthening by insoluble phases (for example thoria in nickel) occurs in the same way as precipitation hardening.

1.2.4   Two-dimensional Defects

The most important two-dimensional defect in solids is the grain boundary. Most solids do not consist of a single crystal but aggregates of crystals of differing orientations and sizes, known as grains. The boundaries of these grains are regions of high energy, since they are in essence internal surfaces, which lends them the power to enhance diffusion rates and to collect particulate foreign matter. Grain boundaries are best avoided in semiconductor devices – they are made from a single crystal. Though the grain boundary is a region of high energy, it is not a source of weakness in a solid; in general, polycrystalline materials do not have lower strengths than single crystals, except for whiskers.

Whiskers are single crystals of very high aspect ratio (that is, long and thin) that possess very high strengths because they contain no cracks, surface steps or dislocations which are the cause of weakness in solids (though some contain a single, axial, screw dislocation, the origin of the nucleation and growth of the whisker in the first place). As far as can be ascertained, any solid which is normally crystalline can be grown in whisker form. In order to be of use, however, ceramic whiskers must be protected by embedding them in a low-yielding matrix – in other words they must be used in reinforced materials. Table 1.1 shows that theoretical yield strengths are almost obtainable in whiskers. The figure of merit for a strong material is its Young's modulus (Y) divided by its density (ρ) which reaches about 300 MNm/kg for graphite whiskers compared with 30 MNm/kg for structural steel. Of course, any single crystal containing immobile dislocations and having no surface defects will exhibit very high strength, but as soon as its surface is slightly damaged it will be susceptible to crack propagation and catastrophic failure at low stress.

Stacking faults are another two-dimensional defect which occur when the sequence of atomic layers in a crystal goes wrong; for example, in a close-packed arrangement the hcp sequence ABABA... can become ABCABA... , or vice versa. The presence of stacking faults in epitaxial

Table 1.1   Mechanical Properties of Whiskers

Material	Tensile Strength (GPa)	Y (GPa)	ρ (kg/m³)	Y/ρ (MNm/kg)
Graphite	21	700	2200	320
SiC	21	493	3000	164
Al_2O_3	21	535	4000	134
Si_3N_4	14	387	3100	125
Si	7.7	183	2300	80
Fe	13	197	7800	25

silicon layers is a useful means of measuring layer thickness and crystallographic orientation (see problem 1.10). In certain substances, such as silicon carbide, ribbon crystals may be grown which are so full of stacking faults that it is hard to say if the various regions of the ribbon are hexagonal or cubic.

1.2.5   Three-dimensional Defects

There are two three-dimensional defects: the precipitate or inclusion of a foreign phase, which is very useful in strengthening metals by anchoring dislocations, and the void, which is an empty space filled by gas. Both types of defect try to lower their surface energies by staying at grain boundaries. Voids shrink by vacancy diffusion, which is quite rapid at grain boundaries, but rather slow in the middle of a grain. Voids are of most importance in materials made by compacting and sintering powders, such as ferrite magnets and transformer cores and the artefacts of powder metallurgy. Both voids and precipitates play a similar role in pinning dislocations and magnetic domain walls.

1.3   Binary Phase Diagrams

1.3.1   Gibbs's Phase Rule

A *phase* is a region of homogeneity in a chemical system – thus a volume of gas is always a single phase because each part of it looks just like any other part: at any point in the gas the surroundings look the same, which is the test for a phase. For the same reason a solution is a single phase, as are miscible liquids and solid solutions. Each differently structured solid constitutes a separate phase – diamond and graphite are both made up solely of carbon atoms, but they are two different phases – because the surroundings of atoms in diamond and graphite are different.

The relationships between phases are displayed on a phase diagram, so these are very useful for preparing materials. Consideration of the conditions for the thermodynamic equilibrium of a system lead to a simple rule, first discovered by J. W. Gibbs, usually called just *the* phase rule, as though there were no others of any account

$$P + F = C + 2 \qquad (1.6)$$

Here, P is the number of phases present, F is the number of degrees of freedom and C is the number of components in the system. F is the number of variables – temperature, pressure, composition etc. – which are independently controllable. A component is the smallest number of elements whose quantities must be specified in order to determine the composition of a system. For example, the system ice, water, steam has three phases, but only one component, because if the amount of hydrogen is specified, the amount of oxygen and water is fixed. The system SrO/Fe_2O_3 has two components, because if, say, the amounts of Sr and Fe are specified, the amount of oxygen is thereby fixed.

With the aid of the phase rule we can construct phase diagrams indicating the regions where the various phases are stable. Very often we shall be concerned with only two-component systems and therefore binary phase diagrams, for which the phase rule may be modified. Because the pressure is usually fixed at one atmosphere, the phase rule becomes

$$P + F' = 3 \qquad (1.7)$$

where F' is the remaining number of degrees of freedom (pressure having been used already) and we have used the fact that $C = 2$. Binary phase diagrams are usually plotted as temperature against composition diagrams with temperature along the vertical axis.

Equation (1.7) shows that three phases can occur only at a point, and two along a line, on the phase diagram. When two solids form a complete range of solid solution there can only be two phases present (we discount the unlikely situation where the temperature is high enough for the liquid to boil off into gas) and the phase diagram takes the form of figure 1.19.

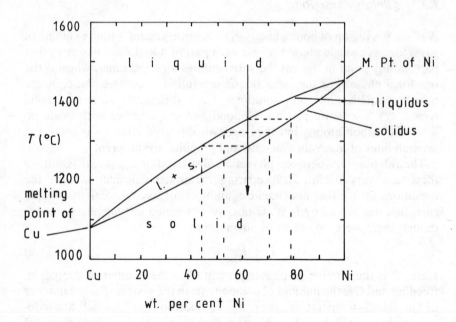

Figure 1.19 *The Ni–Cu phase diagram*

This type of phase diagram obtains when the two constituent solids have similar structures and atomic sizes, such as nickel and copper. In figure 1.19 only liquid lies above the upper line called the *liquidus* and only solid below the lower line known as the *solidus*. Between these two lines lies a two-phase region where liquid and solid may be found together, so that by equation (1.7) there can be only one degree of freedom. Thus if the temperature varies in this region, the composition is fixed and vice versa.

Suppose we mix together Cu and Ni so that the composition is 62 per cent Ni and 38 per cent Cu (weight percentages are normally given, unless stated otherwise). We then melt them together at a temperature well above the liquidus, and cool the liquid down so that it goes through the liquidus at 1350°C and the solidus at 1285°C. Equilibrium conditions are assumed to occur throughout the cooling process (so the cooling must be slow). At 1350°C the first solid separates, whose composition must be 79 per cent Ni, because 1350°C solid of that composition lies on the solidus, while the liquid composition is still 62 per cent Ni, as that is the point on the liquidus corresponding to 1350°C. Further cooling takes the temperature to 1320°C, inside the two-phase region, so the liquid composition must be 53 per cent Ni while the solid is 71 per cent Ni as these are the points on liquidus and solidus corresponding to 1320°C. When cooling has proceeded to 1285°C, the last drop of liquid is left, which has composition 44 per cent Ni, while the solid contains 62 per cent Ni.

During cooling, the solid which separates out of the melt starts off richer in Ni than the original composition, while the liquid is poorer. The first solid that separates out must lose atoms of Ni to the melt as it cools, in order to maintain the solid composition at equilibrium. This requires diffusion in the solid, which may be fairly slow. Fast cooling (quenching) will not allow this diffusion to occur and non-equilibrium solid results. At any temperature in the two-phase region the percentage of each phase may be calculated by means of the lever rule.

1.3.2 The Lever Rule

This useful rule gets its name from an analogy with mechanical levers, which need not be pursued. The material balance of A and B in liquid and solid phases may be worked out to show that at equilibrium and temperature T_2 in figure 1.20, the weight fraction of solid is given by

$$s = (x - w)/(y - w)$$

while the weight fraction of liquid is

$$l = (y - x)/(y - w) = 1 - s$$

We can readily derive these results. The solid contains y per cent B and $(100 - y)$ per cent A, while the liquid contains w per cent B and $(100 - w)$

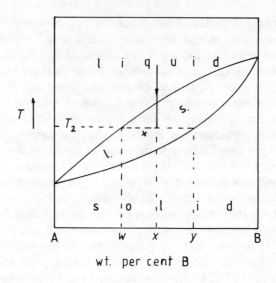

Figure 1.20   *The lever rule*

per cent A. The amount of B in the solid is just sy per cent and in the liquid lw per cent, and these must add up to x per cent:

$$x = sy + lw = sy + (1 - s)w$$

which leads to

$$s = (x - w)/(y - w)$$

As an example, let us consider the copper–nickel phase diagram of figure 1.19. Suppose we have an alloy of copper and nickel containing 30 per cent nickel, which is heated to 1300°C and then cooled to 1200°C, what is the percentage of liquid at the latter temperature? The horizontal intercepts on liquidus and solidus are respectively 23 per cent and 38 per cent, so the lever rule give the percentage of liquid as

$$l = 100 \times (38 - 30)/(38 - 23) = 53 \text{ per cent}$$

Quenching this alloy will yield essentially a two-phase material with differing amounts of nickel in each phase. In principle this will gradually return to equilibrium by diffusion, but the process is very, very slow at room temperature.

1.3.3 Eutectics

Nickel and copper are of similar structure and atomic size, whereas lead and tin are not, so they will not form a continuous range of solid solutions, but regions of limited solid solubility do occur. The phase diagram that results often looks like figure 1.21. Tin is more soluble in lead than vice versa, so the α-phase region is much more extensive than the β. α in this case is just the solid solution of tin in lead, which has the same structure as lead. The tin atoms are randomly distributed throughout the solid – if they showed some predilection for a particular site, then *ordering* is said to have occurred and a new phase would be defined. Alloy phases are usually given Greek letters in order starting from the left-hand side of the binary phase diagram – α, β, γ, δ, ε, ξ, η, θ and κ – more are seldom required (copper–aluminium has this many).

The points of maximum solid solubility are joined by a straight line in figure 1.21, which defines the *eutectic* (from the Greek, meaning easily melted) temperature of 183°C. Below the eutectic temperature, no liquid is found, only a mixture of the two solids, α, and β. If a solid containing 62 per cent by weight of tin (the eutectic composition) is melted and then cooled, it will be found to have a sharp melting point, the eutectic temperature. The eutectic point is invariant as three phases (α, β and liquid) are in equilibrium, and (1.7) shows there can be no degrees of freedom.

Above the eutectic temperature are two two-phase areas where solid of one type or the other is in equilibrium with liquid. For example, a 60/40 solder (60 weight per cent lead and 40 weight per cent tin) will begin to freeze at 245°C, when the first α separates out and will get progressively pastier as the temperature falls until the solder becomes wholly solid at 183°C. The lever rule can be used in any two-phase region to find the

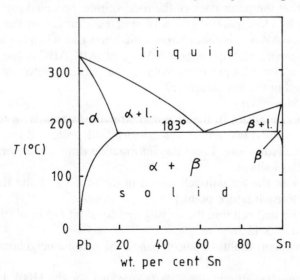

Figure 1.21 *The Pb–Sn phase diagram*

percentage of solid or liquid. Taking the 60/40 solder, we can calculate the amount of liquid at 200°C, say, by drawing a horizontal line and noting the intercepts with the lines to the left and right which define the boundary of the α plus liquid region. These intercepts are at 17 per cent and 57 per cent, so the percentage of liquid by the lever rule is

$$l = 100 \times (40 - 17)/(57 - 17) = 57.5 \text{ per cent}$$

When the eutectic temperature is reached, the liquid remaining must be of the eutectic composition (62 per cent tin) and the percentage is

$$l_e = 100 \times (40 - 19)/(62 - 19) = 49 \text{ per cent}$$

Cooling below the eutectic temperature causes this eutectic liquid to freeze and to transform into a mixture of α and β. Just below 183°, the percentage of α is given by the lever rule as

$$\alpha = 100 \times (97.5 - 40)/(97.5 - 19) = 73 \text{ per cent}$$

Any composition containing between 19 per cent and 97.5 per cent tin will give rise to eutectic liquid, but compositions outside these limits will not. For example, a 10 per cent alloy will first form solid on cooling down at about 300°C, and the last liquid will vanish at 260°C, which is well above the eutectic temperature, and it will contain 30 per cent tin – well below the eutectic composition.

Problems

1.1. Show that the percentage of the total volume occupied (the *packing factor*) by close packed layers of spheres arranged in the stacking sequence AAA... is about 52 per cent. Show that when the layers are in close packed sequences ABAB... or ABCABC... the packing factor is about 74 per cent. What is the packing factor for spheres arranged in the bcc structure?
[Ans. 68 per cent]

1.2. Silicon crystallizes in the diamond structure, which is an fcc lattice with a basis of two atoms, one at each lattice site and one at $\frac{1}{4}, \frac{1}{4}, \frac{1}{4}$ from the lattice site. From this information draw a plan view of the unit cell of silicon.
[Start with the fcc unit cell shown in figure 1.5 and add the silicon atoms to each lattice point.]
From the unit cell find the density and covalent radius of silicon, for which $a = 5.43$ Å.
[The covalent radius of silicon is half the nearest neighbour separation.]
What are the atomic densities (atoms/m²) for the [100], [111] and [110] planes? What is the packing factor for the diamond structure?

[*Ans. 2330 kg/m³, 1.18 Å. 6.8 × 10¹⁸/m², 7.8 × 10¹⁸/m², 9.6 × 10¹⁸/m². 34 per cent*]

1.3. GaAs is a III–V compound semiconductor which is important for high frequency devices. It crystallizes in the zincblende (sometimes called sphalerite) structure, whch has an fcc lattice and a basis of one Ga atom at the lattice sites and one As atom at $\frac{1}{4}, \frac{1}{4}, \frac{1}{4}$ from each lattice site. There are four formula units to the unit cell. Draw a plan view of the unit cell. If the lattice parameter is 5.65 Å, find the density of GaAs. Looking at the unit cell, it is obvious that the [100] planes contain only gallium atoms and no arsenic atoms. Which low-index planes contain only arsenic atoms? Which both?
[*Ans. 5325 kg/m³*]

1.4. Suppose $CaCl_2$ dissolves in KCl so that a Ca^{2+} ion replaces a K^+ ion and the two Cl^- ions from $CaCl_2$ replace one Cl^- ion from the KCl. Assuming that the cations are the same size, what will be the density change when 0.02 mole per cent $CaCl_2$ is added to KCl?
[Consider the two cases where (a) there is no volume change, (b) the volume changes to accommodate the additional Cl^- ions. KCl has the NaCl structure, with $a = 6.293$ Å.]
Now consider the case where one K^+ vacancy forms for each Ca^{2+} ion, while the Ca^{2+} replaces a K^+. The chloride ions from $CaCl_2$ now occupy normal Cl^- sites, causing no change in density. What will then be the change in density when 0.02 mole per cent $CaCl_2$ is added to KCl?
[The observed change in density is -0.16 kg/m³.]
[*Ans. (a) + 0.19 kg/m³; (b) ≈ 0; − 0.20 kg/m³*]

1.5. Use figure 1.9 to find the volume fraction of vacancies in aluminium at its melting point of 933 K. If the energy of vacancy formation is 0.8 eV in aluminium, what contribution do vacancies make to the specific heat at 933 K? Compare this to Dulong and Petit's value of 25 J/mole/K for the specific heat above room temperature.
[Remember $C_v = \partial U/\partial T = E_v \partial N_v/\partial T$.]
[*Ans. 0.9 J/mole/K*]

1.6. A rectangular bar of aluminium $100 \times 20 \times 10$ mm is bent into a circular arc as in figure P1.6. If all the plastic deformation is produced by a static array of edge dislocations as shown in figure P1.6, what would be their density if the atomic planes are 2 Å apart? What is the energy stored in the dislocations? What is the smallest size of F, the constant force required to bend the bar?
[G for Al is 27 GPa and the length of the Burgers vector is 1 Å.]
[*Ans. 2.5 × 10¹⁸/m²; 13.5 kJ; 1.1 MN*]

1.7. A cubic crystal of side length l contains a single, straight, edge dislocation, also of length l. The Burgers vector, b, is perpendicular to one face and a shear stress, σ, is applied to the crystal so that the

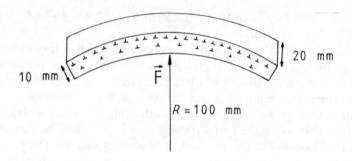

Figure P1.6   *Edge dislocation in a deformed bar*

dislocation is moved completely across it a distance l. If a force, F, is applied to the dislocation in the direction of its Burgers vector, find the work done in moving the dislocation. By comparing this to the work done by the shear stress, show that $f = \sigma \cdot b$, where $f (\equiv F/l)$ is the force per unit length of dislocation.

[This is the formula of Mott and Nabarro.]

1.8.   A metal is strengthened by uniformly dispersing throughout its volume spherical particles of an insoluble second phase of diameter 100 nm. These particles act as pinning sites for dislocations. Suppose the Burgers vector of the dislocations is of length 2 Å and the volume fraction of the dispersed phase is 1 per cent. The maximum shear stress is given by the formula $\sigma \approx Gb/l$, where l is the distance between pinning sites. Show that the maximum shear stress is $G/1870$.

[Find the number of dispersed particles per unit volume, n. The average distance between particles is then $\sqrt[3]{(1/n)}$.]

1.9.   Two identical edge dislocations of opposite sign may meet to form a line of vacancies of the same length as each dislocation. If the interatomic spacing and the length of the Burgers vector are both 2 Å and the energy of vacancy formation is 1 eV, find the energy/unit length of an edge dislocation. If edge and screw dislocations have the same energy, show that the shear modulus of the metal is 10 GPa. Repeat the calculation for $b = 1$ Å. What do you conclude?
[Ans. 40 GPa]

1.10.   In semiconductor device manufacture it is often necessary to deposit a thin layer of, for example, silicon onto a slice of single crystal silicon of opposite conductivity type. This is known as *epitaxy*. The 'grown' layer takes up the same crystallographic orientation as the substrate. When impurities are present at the interface, stacking faults are nucleated which propagate through the layer as shown in figure P1.10(a). The surface of the layer may be etched with an

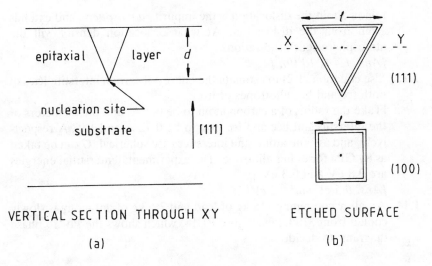

Figure P1.10 *(a) A cross-section of a stacking fault in an epi-layer in Si; (b) etched stacking faults in silicon epi-layers*

aqueous solution of CrO_3 and HF, which delineates the stacking fault as shown in figure P1.10(b). When the crystal surface is a [111] plane the stacking fault occurs over the surface of a regular tetrahedron, one of whose faces is delineated by etching, and one of whose vertices is at the interface. If the layer thickness is d and the side length of the tetrahedron is l, show that $l = 1.255d$. In the case of epi-layers grown on [100] faces, the stacking-fault polyhedron is half of a regular octahedron and the etched surface reveals square etch-figures whose sides are the edges of the regular octahedron, one of whose vertices lies at the interface. In this case show that $l = 1.414d$.

1.11. During severe cold working a bar of copper gets warmer by 20 K, since the work expended appears mostly as heat. About 10 per cent of the work done, however, is stored in dislocations. If the shear modulus of copper is 50 GPa and the length of the Burgers vector is 2 Å, find the dislocation density.
[Specific heat of copper = 400 J/kg/K.]
[Ans. $3.6 \times 10^{15}/m^2$]

1.12. Dissolved impurity atoms are bound to dislocations with a binding energy of about 0.2 eV. If there is equilibrium between the impurity atoms bound to dislocations and those dissolved in the host solid, then the Maxwell–Boltzmann distribution applies. Given that the concentration of an impurity is 1 ppm, what will be the concentration of impurity atoms bound to dislocations at room temperature? The

region about the dislocation is the impurity *atmosphere*, and extends
1 nm from the dislocation. At what dislocation density will the
atmosphere reach saturation?
[*Ans. 1.4 × 10¹⁴/m²*]

1.13. Use equation (1.2) to estimate the energy of carbon interstitialcies in
both fcc and bcc allotropes of iron.
[Take the radius of a carbon atom to be 0.7 Å, and the diameters of
the interstices in bcc and fcc iron to be 0.72 Å and 1.07 Å respectively, and assume atoms and interstices are spherical. *G* can be taken
as 80 GPa for either allotrope. The experimental interstitial energies
are 0.8 eV and 0.3 eV.]
[*Ans. 0.5 eV and 0.2 eV*]

1.14. An alloy containing 115 kg of lead and 7.5 kg of antimony is slowly
cooled from 310°C; use figure P1.14, which shows the Pb–Sb phase
diagram, to decide

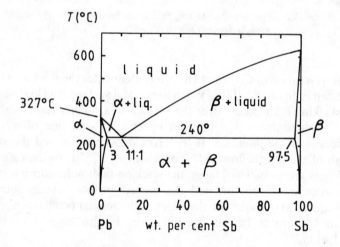

Figure P1.14   *The lead–antimony phase diagram*

(a) the temperature when the first solid begins to separate
(b) the composition of the solid in (a)
(c) the weight fraction of liquid at 241°C
(d) the percentage of α and β solid solutions at 239°C
(e) how much Sb must be added to the alloy to produce a composi-
    tion with a sharp melting point.
[Antimonial lead is used in lead–acid batteries to give greater
strength, to lower the casting temperature and make better casts.
Can you think of any disadvantages in using this alloy?]

[*Ans. (a) 279°C, (b) 1.7 per cent Sb, (c) 0.385, (d) 96.7 per cent α, 3.3 per cent β, (e) 6.9 kg*]

1.15. In the preparation of $SrO.6Fe_2O_3$ it is customary to use less than the required amount of Fe_2O_3 for two reasons: first, to have some eutectic liquid present during sintering (a process which causes densification of a powder compact) at 1210°C, and second, because some iron oxide is picked up during processing of the powdered material. Use figure P1.15 to find the molecular formula which would give 3 w/o of eutectic liquid at 1210°C. In this case, what would be the percentage of $7SrO.5Fe_2O_3$ at 300 K and how much of the $SrO.6Fe_2O_3$ at 300 K would have come from the eutectic liquid? [*Ans. $SrO.5.5Fe_2O_3$, 2.1 per cent and 0.92 per cent*]

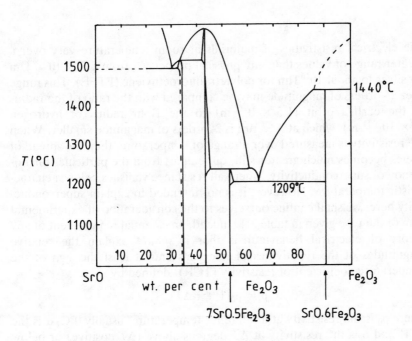

Figure P1.15　*The SrO–Fe₂O₃ phase diagram*

2
The Classical Theory of Electrical Conduction

The electrical resistivities of materials at room temperature vary over a greater range of values than any physical property – from 1.5×10^{-8} Ωm for silver to about 10^{16} Ωm for polytetrafluoroethylene (PTFE). This range over 24 orders of magnitude may be compared with the ratio of the radius of the earth's orbit $(1.5 \times 10^{11}$ m$)$ to the Bohr radius of hydrogen $(5 \times 10^{-11}$ m$)$, which at 3×10^{21} is $2\frac{1}{2}$ orders of magnitude smaller. When the resistivity is measured over a range of temperature, then the spread of values becomes much greater still, quite apart from the particular phenomenon of superconductivity, where all resistance vanishes below a characteristic temperature. However, it is not intended to explain superconductivity here; we shall confine ourselves to the consideration of experimental data of the type given in table 2.1. Indeed, an essential requirement of any theory of electrical behaviour is that it *should* explain the relative magnitudes of the resistivities of materials and at least the sign of the temperature coefficient of resistivity (TCR), defined by

$$\rho = \rho_0(1 + \alpha\Delta T)$$

where ρ_0 is the resistivity at a reference temperature, usually 0°C, α is the TCR and ρ is the resistivity at ΔT degrees above (ΔT positive) or below (ΔT negative) the reference temperature.

Examination of table 2.1 shows that the pure metals are of low, and given the vast range of values for all materials, constant resistivity with a positive TCR of about 0.004/K at room temperature. Alloys have resistivities one or two orders of magnitude greater than pure metals, with lower, but positive TCRs. Non-metals appear to have very variable resistivities, tending to higher values, often with large, negative TCRs. It has been customary to say that those materials with very high resistivities are insulators and those with resistivities intermediate between metals and insulators are semiconductors, both having large, negative TCRs. A glance

Table 2.1 Resistivities and TCRs at 293 K

Material	ρ (Ω m)	TCR (%/K)
Silver	1.6×10^{-8}	+ 0.41
Copper	1.7×10^{-8}	+ 0.43
Aluminium	2.7×10^{-8}	+ 0.43
Sodium	5.0×10^{-8}	+ 0.4
Tungsten	5.7×10^{-8}	+ 0.45
Iron	9.7×10^{-8}	+ 0.5
Platinum	10.5×10^{-8}	+ 0.39
Tantalum	13.5×10^{-8}	+ 0.38
Manganin (87Cul3Mn)	38×10^{-8}	+ 0.001
Constantan (57CU43Ni)	49×10^{-8}	+ 0.002
Nichrome (80Ni20Cr)	112×10^{-8}	+ 0.0085
SiC, commercial	$1-2 \times 10^{-6}$	− 0.15
Graphite, commercial	$c.\ 1 \times 10^{-5}$	− 0.07
InAs, very pure	3×10^{-3}	− 1.7
Tellurium, very pure	4×10^{-3}	− 2
Germanium, diode grade	1×10^{-3}	+ 0.4
Germanium, very pure	5×10^{-1}	− 4
Silicon, transistor grade	1×10^{-1}	+ 0.8
Silicon, very pure	1×10^{-3}	− 7
Anthracene	3	− 10
Selenium, amorphous	$c.\ 1 \times 10^{10}$	− 15
Silica	$c.\ 1 \times 10^{13}$	negative
Alumina	$c.\ 1 \times 10^{14}$	negative
Sulphur	$c.\ 1 \times 10^{15}$	negative
PTFE	$c.\ 1 \times 10^{16}$	negative

at table 2.1 shows that this is not the case: silicon carbide is truly a semiconductor, yet it appears to have a resistivity no more than that of nichrome, while several device-grade semiconductors have positive TCRs. The reasons for the traditional divisions of materials are as follows:

(1) When a sample of fairly pure metal is prepared – say 99.9 per cent – we can predict the resistivity at room temperature and its TCR very well, and increasing the purity to 99.99 per cent or 99.999 per cent will not affect this much. The same goes for alloys, but more so – an 80/20 Ni–Cr alloy will have very nearly the same resistivity and TCR as a 79/21 nichrome.

(2) When a sample of, say, sulphur is prepared which is 99.9 per cent pure, we can reasonably be sure it will have a large negative TCR (though it may prove difficult to measure with accuracy), and its resistivity will be very large – so large, in fact, that it is hard to measure because of problems with technique (surface leakage is unpredictable, for

example). Making the sulphur 99.999 per cent pure would not really have much effect on the resistivity measured or its TCR. Sulphur is an insulator.

(3) When the resistivity of a silicon sample is measured, its value is found to be dependent, not so much on the purity of the sample, as on the nature of the impurities. The TCR is also found to be very variable, perhaps positive, perhaps negative. And a high resistivity may be found in samples which are not necessarily of very high purity: the only certainty is that a low resistivity indicates some sort of electrically active impurity is present.

These types of behaviour have come to be known as 'metallic', 'insulating' and 'semiconducting', respectively. We shall strive to explain all three types, starting with metallic conduction.

2.1 Drude's Free Electron Theory

J. J. Thomson discovered the electron in 1890, and in 1900 P. K. Drude published (*Ann. Physik*, **1**, 566) a theory of electrical conduction which has formed the basis for subsequent theories on the subject. Drude's idea was very simple: metals contained electrons which were not, as in other materials, bound to a particular atom, but were free to wander through an assemblage of cations which made up the metal's lattice. Insulators, possessing few, if any, free electrons could not conduct electricity. In classical theory, free electrons are taken to be particles inside the body of the metal obeying Maxwell–Boltzmann statistics and Newton's laws of motion. A number of important phenomena may be quantitatively explained by these assumptions, as we shall see.

2.1.1 The Drift Velocity, Mobility and Ohm's Law

Suppose an electric field, $\mathscr{E}$, is applied to a metal in which there are n free electrons per unit volume, each bearing a charge of $-e$ coulombs. The force, F, acting on the electron due to the field is $-e\mathscr{E}$ and the equation of motion for a single electron is

$$F = -e\mathscr{E} = ma \tag{2.1}$$

where m is the mass of an electron and a is its acceleration. The acceleration is $-e\mathscr{E}/m$, or $1.75 \times 10^{11}\mathscr{E}$ m/s so the electron would reach a speed equal to 1 per cent of the speed of light in a field of 1 V/m in about 5 μs without the intervention of another force. Now it seems reasonable to expect that the free electrons will bump into a cation of the lattice from time to time: suppose the average time between such collisions were τ

seconds. Immediately after a collision we further suppose that the velocity of the electron averages to zero; that is to say, the electron has no 'memory' of the momentum acquired from the field and that its thermal velocity averages to zero. In a time of τ s the electron will attain a velocity given by

$$v_d = a\tau \tag{2.2}$$

v_d is called the drift velocity of the electron. Substituting in (2.2) for a, we can find the drift velocity:

$$v_d = - e\mathscr{E}\tau/m \tag{2.3}$$

Usually this is re-written in the form

$$v_d = - \mu\mathscr{E}$$

where μ ($\equiv e\tau/m$) is the *mobility* of the electron, or its drift velocity in unit field. Considering unit volume of the metal, which contains n free electrons, each of charge $- e$, the total charge is $- ne$, so the charge crossing unit area in unit time must be $- nev_d$, which is the current density, J (in A/m^2):

$$J = - nev_d \tag{2.4}$$

Substituting for v_d from (2.3) yields

$$J = ne^2\mathscr{E}\tau/m \tag{2.5}$$

or

$$J = \sigma\mathscr{E} \tag{2.6}$$

where σ ($= 1/\rho$) is the conductivity of the metal. Comparing (2.5) with (2.6) gives

$$\sigma = ne^2\tau/m \tag{2.7}$$

which is

$$\sigma = ne\mu \tag{2.8}$$

Now the resistivity is defined by

$$R = \rho l/A \tag{2.9}$$

where R is the resistance of a specimen of length l and cross-sectional area A (see figure 2.1), so as $\sigma = 1/\rho$ and $\mathscr{E} = V/l$, equation (2.6) gives

$$J = V/\rho l \tag{2.10}$$

From (2.9) $\rho l = RA$, thus (2.10) becomes

$$J = V/RA = I/A$$

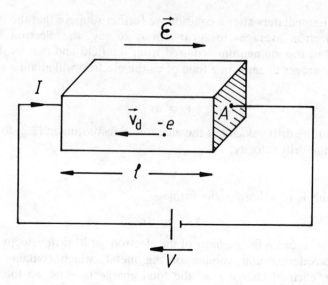

Figure 2.1 *Electron drift in a metal bar under the influence of an electric field*

yielding

$$V = IR \qquad (2.11)$$

This is Ohm's law in customary guise. Equation (2.6) is a form more suitable for solving electromagnetic problems and was derived from a simple free-electron model of a conductor in which lattice collisions restrict the speeds attained by the electrons.

2.1.2 An Estimate of Mobility and Drift Velocity

There is one free electron per atom in copper (it is sure to be one or two since the valency is one or two). The Periodic Table given at the end of the book shows that the unit cell of copper is fcc and has lattice parameter 3.6 Å, so that there are $4/(3.6 \times 10^{-10})^3$ atoms/m^3, or 8.5×10^{28} free electrons/m^3. At room temperature the resistivity of copper is 1.67×10^{-8} Ωm, so the conductivity is 6×10^7 S/m. Thus, (2.8) gives

$$\mu = \sigma/ne$$
$$= 6 \times 10^7/(8.5 \times 10^{28} \times 1.6 \times 10^{-19})$$
$$= 4 \times 10^{-3} \text{ m}^2/\text{V/s}$$

The mobility is not great: for example, in a typical house a 15 A wire, 3 m long, will have a resistance of about 0.03 Ω, so the voltage drop at maximum current is 0.45 V and the field is 0.15 V/m. Thus the drift velocity would be only $4 \times 10^{-3} \times 0.15$ m/s, or about $\frac{1}{2}$ mm/s. The average thermal speed of electrons in copper is found by allocating $\frac{1}{2}k_BT$ joules of energy to each degree of freedom available to an electron. There can only be three translational degrees of freedom for a 'monatomic' electron gas, so that the average thermal energy is $1\frac{1}{2}k_BT$ and this is equal to the average translational energy.

$$1\tfrac{1}{2}k_BT = \tfrac{1}{2}mv_{th}{}^2$$
$$v_{th} = \surd(3k_BT/m) \tag{2.12}$$

Taking T to be 300 K, (2.12) yields

$$v_{th} = \surd(3 \times 1.38 \times 10^{-23} \times 300/9.1 \times 10^{-31})$$
$$\approx 10^5 \text{ m/s}$$

In our household conductor the ratio of thermal to drift velocities is about $10^8 : 1$. Even in a tungsten filament in a light bulb, where the electron has a relatively large drift velocity (see problem 2.1), the ratio is still very large.

2.1.3 The Mean Free Path

The average distance travelled by an electron between collisions is its mean free path, l_m, given by

$$l_m = \tau<v> \tag{2.13}$$

where $<v>$ is the average velocity of the electrons, made up of the thermal and drift velocities. Under normal conditions, the thermal velocity is, as we have seen, much greater than the drift velocity and (2.13) may be written

$$l_m = \tau v_{th}$$

Now τ can be found from

$$\mu = e\tau/m$$
$$\tau = \mu m/e$$
$$\tau = 4 \times 10^{-3} \times 9.1 \times 10^{-31}/1.6 \times 10^{-19}$$
$$= 2.3 \times 10^{-14} \text{ s}$$

Substituting in (2.13) and taking $<v>$ as 10^5 m/s, we find

$$l_m = 2.3 \times 10^{-14} \times 10^5 = 2.3 \text{ nm}$$

which may be compared with the 0.2 nm separating atoms in copper. The electron appears to travel about ten times the average distance between atoms before colliding with one. In the next chapter we shall see that l_m is much greater than this, though the drift velocity remains the same.

2.2 The Hall Effect

Electrons in motion in a magnetic field are subject to a force usually known as the Lorentz force

$$F = - ev \times B \qquad (2.14)$$

where B is the magnetic induction. The direction of the force is given by the right-hand corkscrew rule, but we must bear in mind the minus sign due to the negative charge on the electron. Figure 2.2 shows a slab of material with a battery connected across it so that the electrons flow from right to left (conventional current from left to right). If a magnetic field is applied perpendicular to the electric field, then the Lorentz force acts upwards on the electrons, pushing them to the top of the sample. Here they accumulate and produce a field $\mathscr{E}_H$, known as the Hall field, which acts on the electrons so as to oppose the Lorentz force. The magnitudes of the two forces must be equal at equilibrium, so

$$e\mathscr{E}_H = evB \qquad (2.15)$$

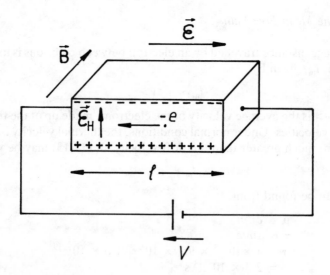

Figure 2.2 *The Hall field in a bar of material*

The magnitude of the vector product $v \times B$ must be vB because v and B are perpendicular. The three vectors, $\mathscr{E}$, $\mathscr{E}_H$ and B are mutually perpendicular. Now, the v in (2.15) is the electron's instantaneous velocity, so that the average value of the Lorentz force is to be found from an average value of v, given by

$$<v> = <v_{th} + v_d>$$

if

$$<v_{th} + v_d> = <v_{th}> + <v_d>$$

and $<v_{th}>$ is zero, as thermal velocities are random in direction, then $<v> = <v_d>$, so that (2.15) becomes

$$\mathscr{E}_H = v_d B \qquad (2.16)$$

If v_d is of the order of 1–10 mm/s and taking a reasonable value for B of 0.5 T (5000 G), we can see that the Hall field in a sample of copper will be a few mV/m, which is a rather small quantity. In (2.16) v_d may be replaced by $\mu\mathscr{E}$ to give

$$\mathscr{E}_H = \mu\mathscr{E}B \qquad (2.17)$$

The Hall field is the product of electric field, magnetic induction and the mobility of the electrons in the material. And in terms of electron concentration and current density, since $v_d = -J/ne$, the magnitude of the Hall field is given by

$$\mathscr{E}_H = (-1/ne)JB \qquad (2.18)$$

$$= R_H JB \qquad (2.19)$$

R_H is the Hall constant of the material, which may be found by a simple measurement. Looking at (2.18) and (2.19) it can be seen that

$$R_H ne = -1 \qquad (2.20)$$

Table 2.2 $R_H ne$ Values for some Metals

Metal	$R_H ne$
Li	− 1.15
Na	− 1.05
K	− 1.08
Rb	− 1.05
Cs	− 1.04
Cu	− 0.68
Ag	− 0.8
Au	− 0.69
Pd	− 0.73
Pt	− 0.21
Cd	+ 0.5
W	+ 1.2
Be	+ 5.0

Table 2.2 shows some values for the product $R_H ne$ from which it is plain that, although (2.20) is true for some metals, it is completely wrong for others, and even has the wrong sign in some cases, notably for beryllium. Now, it may be possible to fudge the magnitude of the Hall constant, but it is out of the question to fudge the sign: the only way out of the impasse is to accept that there must be positive charge carriers in some metals. In 1931 A. H. Wilson suggested that in some materials there were missing electrons, called *holes*, which behaved just like positively charged electrons. Wilson developed his ideas fully in a book entitled *The Theory of Metals* in 1936. For the moment, the existence of positive charge carriers in metals and semiconductors will have to be accepted as an established experimental fact.

2.2.1   The Hall Probe

Equation (2.17) shows that, for given $\mathscr{E}$ and B across a conductor, the size of the Hall field is proportional to the mobility of the charge carriers in the material. For this reason, indium antimonide (InSb), a III–V compound semiconductor, has been popular for use in a small probe designed to measure magnetic fields at a point in space, since it has a very high electron mobility – about 8 m²/V/s, some 2000 times the mobility in copper. Let us suppose we have a small single crystal of indium antimonide about $5 \times 5 \times 1$ mm, which is disposed as in figure 2.3, with a potential of 1 V

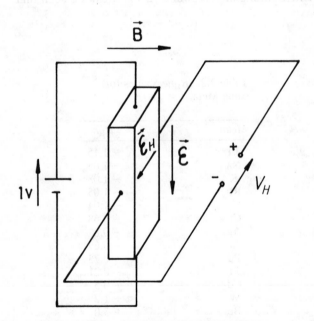

Figure 2.3   *The active element in a Hall probe*

across one pair of faces to produce a current in the semiconductor. The electric field is therefore $1/5 \times 10^{-3}$ V/m, so that the Hall field is

$$\mathscr{E}_H = 8 \times 200 \times B$$

If $B = \mu_0 H$ (μ_0 is the magnetic constant, $4\pi \times 10^{-7}$) in the probe material, which is the case for a non-ferromagnet like InSb, then

$$\mathscr{E}_H = 1600\mu_0 H = 2 \times 10^{-3}H$$

and the Hall voltage is just $5 \times 10^{-3} \times \mathscr{E}_H$, or $10^{-5}H$ volts: a magnetic field of only 1000 A/m (12 Oe; for comparison, the earth's rather weak magnetic field has a magnitude of about $\frac{1}{2}$ Oe) will produce the relatively large Hall voltage of 10 mV. We shall now look at further properties of conductors which appear to depend on the motion of electrons.

2.3 The Wiedemann–Franz Law

In 1853 Wiedemann and Franz stated that for all metals, at a constant and not-too-low temperature, the ratio of electrical and thermal conductivities is constant. On the assumption that electrons alone are responsible for both thermal and electrical conduction in metals, this seems reasonable. If the free electrons in a metal behave like an ideal gas then the thermal conductivity is given by

$$\kappa = \tfrac{1}{3}C_v l_m v_{th} \tag{2.21}$$

Equation (2.21) is a standard result from the kinetic theory of gases, in which C_v is the specific heat at constant volume, l_m the mean free path and v_{th} the thermal speed of the electrons. C_v may be taken as $1\frac{1}{2}nk_B$ for a 'monatomic' ideal gas, so that (2.21) becomes

$$\kappa = \tfrac{1}{2}nk_B l_m v_{th} \tag{2.22}$$

The electrical conductivity is given by (2.7)

$$\sigma = ne^2\tau/m$$

Here, the time between collisions, defined by (2.13), is usually redefined to introduce a factor of two to give better agreement with experiment. This derivation of the Wiedemann–Franz law says that the drift velocity acquired between collisions should be the average of zero and the final velocity, v_d, given in equation (2.2). The collision time is then $\frac{1}{2}l_m/v_{th}$ and the ratio of conductivities is

$$\kappa/\sigma = mk_B v_{th}^2/e^2 \tag{2.23}$$

The thermal speed of the electrons is given by (2.12), so that (2.23) becomes

$$\kappa/\sigma = 3k_B^2 T/e^2 \tag{2.24}$$

Dividing (2.24) by T will clearly give a constant quantity, which must be the same for all metals, known as the Lorenz* number, $\mathcal{L}$:

$$\mathcal{L} = \kappa/\sigma T = 3k_B^2/e^2 \qquad (2.25)$$

Lorenz discovered the constancy of the ratio $\kappa/\sigma T$ experimentally in 1872.

Equation (2.25) is the outstanding achievement of the Drude–Lorenz free electron theory: its validity is illustrated by the Lorenz numbers given in tables 2.3 and 2.4. $3k_B^2/e^2$ has a value of $2.23 \times 10^{-8} \, \text{W}\Omega/\text{K}^2$, and while the tables indicate that $\mathcal{L}$ should be rather higher than this, nevertheless the near-constancy of $\kappa/\sigma T$ over a wide range of metals and temperatures appears as striking proof of the power of the free electron theory of metals.

When the temperature is too much below room temperature, equation (2.25) no longer applies, nor does it hold for a few metals and alloys of low electrical conductivity: for instance, the higher Lorenz numbers of constantan and stainless steel. In these alloys the lattice contribution (that is, the vibration of the atoms in the lattice) to the thermal conductivity is

Table 2.3   Lorenz Numbers for some Metals and Alloys at 293 K

Metal or Alloy	$\mathcal{L}(= \kappa/\sigma T)$ $\times 10^{-8} \, W \, \Omega/K^2$
Aluminium	2.18
Cadmium	2.26
Copper	2.30
Indium	2.34
Lead	2.49
Lithium	2.48
Magnesium	2.38
Molybdenum	2.41
Nickel	2.15
Niobium	2.29
Palladium	2.60
Potassium	2.16
Rhodium	2.27
Rubidium	2.44
Tantalum	2.40
Tin	2.47
Uranium	2.77
Stainless Steel (18/8)	3.33
Phosphor Bronze (1.25%)	2.46
Yellow Brass	2.50
Constantan (55Cu45Ni)	3.56

* L. V. Lorenz was a Danish physicist, not to be confused with H. A. Lorentz, the Dutch physicist. They were responsible for discovering the Lorentz–Lorenz relationship, an important result in optical physics.

Table 2.4   Variation of $\mathscr{L}$ with Temperature for Platinum

T (K)	50	100	150	200	300	500
κ (W/m/K)	118	790	762	748	730	719
σ ($\times 10^7$ S/m)	14.3	3.53	2.04	1.44	0.92	0.54
$\mathscr{L}$ ($\times 10^{-8}$ W Ω/K^2)	1.65	2.24	2.49	2.60	2.65	2.66

becoming appreciable. A quantum-mechanical derivation of $\mathscr{L}$ using Fermi–Dirac statistics makes it $\frac{1}{3}\pi^2 k_B^2/e^2$, which is 2.45×10^{-8} WΩ/K^2, achieving perhaps better agreement with the experimental values.

2.4 Matthiessen's Rule

In deriving Ohm's Law from consideration of the motion of free electrons in an applied electric field it was said that the electrons must collide every so often with something in the metal, such as an atom in the lattice. These collisions should occur frequently enough to allow an average drift velocity to be calculated. Experimental measurements on metallic resistivities at low temperatures indicate that more is involved than just one collision process.

The resistivity of a metal decreases more or less linearly as the temperature is reduced, eventually flattening out at quite low temperatures to a constant residual resistivity. Matthiessen's rule states that the resistivity of a metal containing small amounts of impurities or lattice defects is given by

$$\rho_r = \rho_0 + \rho_L$$

where ρ_0 is the residual resistivity which is independent of temperature and roughly proportional to the amount of impurities present. ρ_L is a temperature-dependent term which approaches zero at 0 K and rises approximately linearly above a characteristic temperature for each metal (the *Debye Temperature*, Θ).

Matthiessen's rule is important because it leads to the idea that, in an ideally pure metal, the resistivity is due solely to the thermal vibrations of the lattice. Quantum theory says that the lattice vibrations must be in discrete packets of energy known as phonons, and the electrons are said to be scattered by phonons. Once the thermal vibrations are fully excited (above Θ) the numbers of phonons will be proportional to the temperature, and so will be the resistivity. However, to stick to classical ideas, we can say that the atoms will vibrate with greater amplitude as the temperature rises so that the electrons are more likely to be scattered.

ρ_0 is due to impurities dissolved in the metal (that is, in solid solution), to vacancies – each of which is roughly as good a single impurity atom at scattering electrons – and to a much lesser extent to grain boundaries, dislocations and precipitates. Vacancies are always present, even in well-annealed samples and may contribute significantly to the residual resistivities of very pure metals. Changing the temperature has little effect on the scattering caused by these lattice imperfections.

We can attribute a characteristic time, τ_i, between collisions for an electron and each type of impurity, vacancy, dislocation etc. (often known as the relaxation time). Then the number of collisions/s for the ith impurity is $1/\tau_i$ and the total number of collisions/s with impurities is $\Sigma\, 1/\tau_i$, so we can write

$$1/\tau_T = \Sigma 1/\tau_i + 1/\tau_L \qquad (2.26)$$

where τ_T is the overall relaxation time and τ_L is the relaxation time for lattice scattering. However

$$\sigma = ne^2\tau/m$$

or

$$\rho = m/ne^2\tau$$

so that

$$1/\tau = ne^2\rho/m$$

Thus equation (2.26) becomes, after cancelling the common factor, ne^2/m

$$\rho_T = \Sigma\rho_i + \rho_L$$

Calling $\Sigma\rho_i$, ρ_0, gives

$$\rho_T = \rho_0 + \rho_L$$

which is Matthiessen's rule

The purity of a metal sample may be expressed in terms of the ratio of its resistivity at room temperature (293 K) and its resistivity at liquid helium temperature (4.2 K). The latter is really just the residual resistivity, so the lower it is the higher is the resistivity ratio and the purer is the sample. Resistivity ratios for 'pure' metals vary widely, as table 2.5 shows. Although the resistivity ratio may be a good yardstick for measuring the purity, it must be used with care since only impurities in solid solution scatter electrons well: precipitates make only a small contribution – see problem 2.2. Figure 2.4 shows the resistivities of two samples of the same metal at low temperatures, and we can see that sample A should be roughly twice as pure as sample B provided the impurities are in solid solution.

Platinum is used to make highly accurate and cheap (thin-film types cost only £3 in 1989) thermometers. To make a thin-film thermometer, a layer

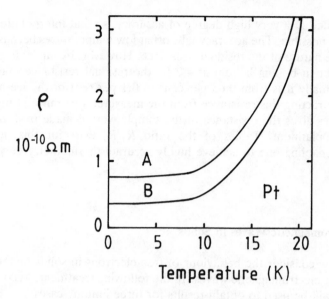

Figure 2.4 *The residual resistivity of two samples of Pt*

Table 2.5   Resistivity Ratios for
High Purity Metals

Metal	$\rho_{293}/\rho_{4.2}$
Tungsten	90 000
Rhenium	45 000
Aluminium	40 000
Molybdenum	14 000
Tantalum	7 000
Gold	2 000
Niobium	2 000
Platinum	2 000
Vanadium	300
Zirconium	200

of platinum is deposited onto an alumina substrate and chemically etched
into a zig-zag pattern to increase its resistance to about 100 Ω or 500 Ω.
The temperature of the film is then held constant at some precisely known
value, such as the ice-point, while a laser beam trims the resistance to
precisely 100 or 500 Ω (usually to 1 part in 10^4). The thin-film thermometer
may then be used at room temperature and above – perhaps to 1000 K.
Standard tables of the resistance ratio R_T/R_0, where R_T is the resistance of
the thermometer at temperature T and R_0 is its resistance at 0°C, have

been compiled to a very high degree of accuracy, so that interpolation to 0.0001 K is possible. The accuracy falls off at low temperatures because of the variable nature of the residual resistance. However, by immersing the thermometer in boiling helium at 4.2 K, the residual resistance may be found (normally it is from 0.05 per cent to 0.1 per cent of the ice-point value). Subtracting this resistance from the measured resistance at higher temperatures gives the resistance of the sample were it made from pure, defect-free platinum. Tables of the ratio R_T/R_0 exist for this 'ideal' platinum, enabling one to achieve highly accurate thermometry down to about 14 K.

2.5   Electromagnetic Waves in Solids

We can now consider the behaviour of free electrons in solids under the influence of electromagnetic waves. In the following treatment, Maxwell's equations will be used to obtain results for three limiting cases:

(1) Where the frequency is 'low' and the electrical conductivity is zero. This corresponds to a lossless or ideal dielectric.
(2) Where the frequency is 'low' and the conductivity is high. This corresponds to a free-electron metal.
(3) Where the frequency of the electromagnetic radiation is 'high' and the conductivity is high. This corresponds to metals at microwave frequencies and above.

By solving Maxwell's equations and making appropriate approximations for each of the three cases listed above, we can predict what will happen to an electromagnetic wave in both dielectrics and metals.

2.5.1   'Low' Frequencies

First let us consider frequencies low enough to allow electron scattering to reach equilibrium. This will be the case when the period of the electromagnetic wave ($1/f$, where f is the frequency in Hz) is much greater than the average time between collisions, τ. That is, when $1/f \gg \tau$, $f\tau \ll 1$, or $\omega\tau \ll 1$, which is the usual form (ω is the angular frequency $2\pi f$). Since the average time between collisions in metals at room temperature is of the order of 10^{-14} s, 'low frequency' really means below microwave frequencies. If you find the next part too mathematical you can skip it and just take note of the results of the analysis, which are embodied in equations (2.35), (2.37), (2.43) and (2.44).

Maxwell's equations are:

$$\nabla \times \boldsymbol{H} = \boldsymbol{J} + \partial \boldsymbol{D}/\partial t \tag{2.27}$$

$$\nabla \times \mathscr{E} = - \partial \boldsymbol{B}/\partial t \tag{2.28}$$

$$\nabla . \boldsymbol{B} = 0$$

$$\left. \begin{array}{l} \nabla . \boldsymbol{D} = q_{\mathrm{v}} \\ \boldsymbol{J} = \sigma \mathscr{E} \\ \boldsymbol{D} = \epsilon \mathscr{E} \\ \boldsymbol{B} = \mu \boldsymbol{H} \end{array} \right\} \tag{2.29}$$

where $\boldsymbol{H}$ is the magnetic field in A/m, $\boldsymbol{J}$ is the current density in A/m^2, $\boldsymbol{D}$ is the electric displacement, $\mathscr{E}$ is the electric field, $\boldsymbol{B}$ is the magnetic induction, q_{v} is the charge/m^3, σ is the electrical conductivity and μ is the magnetic permeability. σ, ϵ and μ are properties which depend on the type of material. Consider a wave travelling along the x-axis as in figure 2.5, with its electric field varying sinusoidally along the y-direction only, so that $\mathscr{E}_x$ and $\mathscr{E}_z$ are both zero. Now as $\mathscr{E}$ varies spatially only in the direction of propagation, $\partial/\partial y$ and $\partial/\partial z$ are also zero. Thus

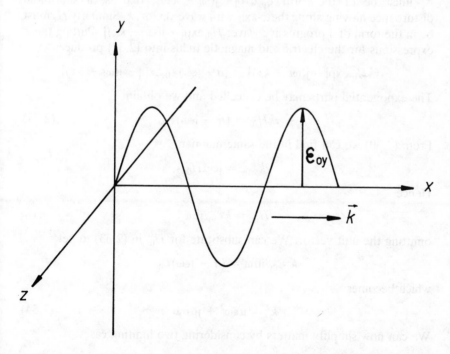

Figure 2.5 *The electric field in an electromagnetic wave propagating along the x-axis*

$$\nabla \times \mathscr{E} = \begin{vmatrix} \hat{x} & \hat{y} & \hat{z} \\ \partial/\partial x & \partial/\partial y & \partial/\partial z \\ \mathscr{E}_x & \mathscr{E}_y & \mathscr{E}_z \end{vmatrix} = \begin{vmatrix} \hat{x} & \hat{y} & \hat{z} \\ \partial/\partial x & 0 & 0 \\ 0 & \mathscr{E}_y & 0 \end{vmatrix}$$

$$= \hat{z}\partial\mathscr{E}_y/\partial x = -\mu\partial H_z/\partial t \tag{2.30}$$

The unit vector, $\hat{z}$, means that H can have only a z-component. $\nabla \times H$ may next be found:

$$\nabla \times H = \begin{vmatrix} \hat{x} & \hat{y} & \hat{z} \\ \partial/\partial x & \partial/\partial y & \partial/\partial z \\ H_x & H_y & H_z \end{vmatrix} = \begin{vmatrix} \hat{x} & \hat{y} & \hat{z} \\ \partial/\partial x & 0 & 0 \\ 0 & 0 & H_z \end{vmatrix}$$

$$= -\hat{y}\partial H_z/\partial x$$

Again, $\partial/\partial y$ and $\partial/\partial z$ are both zero as the wave propagates along the x-axis. Substituting for $\nabla \times H$ in (2.27) yields

$$-\hat{y}\partial H_z/\partial x = J_y + \epsilon\partial\mathscr{E}_y/\partial t \tag{2.31}$$

From (2.29), $J_y = \sigma\mathscr{E}_y$, so (2.31) becomes

$$-\hat{y}\partial H_z/\partial x = (\sigma + \epsilon\partial/\partial t)\mathscr{E}_y \tag{2.32}$$

$\mathscr{E}_y$ must be of the form $\mathscr{E}_{0y}\exp[-j(\omega t - kx)]$, that is, a sinusoidal disturbance moving along the x-axis with wave vector, k. Similarly H_z must be in the form of a progressive wave: $H_{0z}\exp[-j(\omega t - kx)]$. Putting these expressions for the electric and magnetic fields into (2.32) produces

$$-\hat{y}kH_{0z}\exp[-j(\omega t - kx)] = (\sigma - j\epsilon\omega)\mathscr{E}_{0y}\exp[-j(\omega t - kx)]$$

The exponential parts may be cancelled and we obtain

$$-\hat{y}kH_{0z} = (\sigma - j\epsilon\omega)\mathscr{E}_{0y} \tag{2.33}$$

From (2.30) we can find in the same manner

$$\hat{z}k\mathscr{E}_{0y} = j\omega\mu H_{0z}$$

Hence

$$H_{0z} = k\mathscr{E}_{0y}/j\mu\omega$$

omitting the unit vector. We can substitute for H_{0z} in (2.33) to get

$$k^2\mathscr{E}_{0y}/j\mu\omega = (\sigma - j\epsilon\omega)\mathscr{E}_{0y}$$

which becomes

$$k^2 = \mu\epsilon\omega^2 + j\mu\sigma\omega \tag{2.34}$$

We can now simplify matters by considering two limiting cases:

(1) $\sigma = 0$, the medium is non-conducting: a dielectric
(2) σ is large: a metal.

In the first case, for a non-conducting medium, (2.34) would become

$$k^2 = \mu\epsilon\omega^2$$

so that

$$k = \omega\sqrt{(\mu\epsilon)}$$

But the group velocity of the wave is given by $d\omega/dk$, which is the velocity at which it transports energy, or its propagation velocity. And

$$d\omega/dk = v_g = 1/\sqrt{(\mu\epsilon)} \tag{2.35}$$

In this case the medium is said to be non-dispersive because the group velocity is independent of frequency. Were the medium to be a vacuum, for which $\mu = \mu_0$ and $\epsilon = \epsilon_0$, the group velocity would be

$$d\omega/dk = 1/\sqrt{(\mu_0\epsilon_0)} = c$$

This is an expected result, since it shows that the speed of propagation of electromagnetic waves in a vacuum is equal to the speed of light in a vacuum.

In a typical coaxial cable the dielectric is often solid polyethylene which has $\mu = \mu_0$ and a dielectric constant (or relative permittivity, ϵ_r) of 2.3, so that $\epsilon = \epsilon_r\epsilon_0 = 2.3\epsilon_0$ and the wave then propagates at a speed given by

$$v = 1/\sqrt{(2.3\epsilon_0\mu_0)} = 2 \times 10^8 \text{ m/s}$$

2.5.2 The Skin Depth

When the medium is conducting, the group velocity depends on frequency. Thus the medium is dispersive and the wave vector is complex. Let us write

$$k = K\exp(j\phi)$$

where K is the magnitude of the wave vector and ϕ is its argument, so that

$$k^2 = K^2\exp(j2\phi) = \mu\epsilon\omega^2 + j\mu\sigma\omega \tag{2.36}$$

from (2.34). Thus

$$K^2 = \sqrt{([\mu\epsilon\omega^2]^2 + [\mu\sigma\omega]^2)}$$

and

$$2\phi = \tan^{-1}(\sigma/\epsilon\omega)$$

This result may be simplified if $\sigma \gg \epsilon\omega$ (which is almost always true for metals, since σ is about 10^7 S/m and $\epsilon = \epsilon_0$, which implies that ω should be $\ll 10^{18}$). In this case (2.36) becomes

$$k^2 = j\mu\sigma\omega$$

which means that $K^2 = \mu\sigma\omega$ and $2\phi = \pi/2$, giving

$$k = \sqrt{(\mu\sigma\omega)}\exp(j\pi/4)$$
$$= \sqrt{(\mu\sigma\omega)}(\cos 45° + j\sin 45°)$$
$$= \sqrt{(\mu\sigma\omega/2)} + j\sqrt{(\mu\sigma\omega/2)}$$

The imaginary part of the wave vector gives rise to attenuation of the field, because

$$\mathscr{E}_y = \mathscr{E}_{0y}\exp[-j(\omega t - kx)]$$

which becomes

$$\mathscr{E}_y = \mathscr{E}_{0y}\exp[-j(\omega t - ax)]\exp(-bx)$$

where $a \equiv \sqrt{(\mu\sigma\omega/2)}$ and $b = \sqrt{(\mu\sigma\omega/2)}$. The attenuating part, $\exp(-bx)$, causes the wave amplitude to be reduced to $1/e$ of its initial value in a distance given by

$$\delta \equiv 1/b = \sqrt{(2/\mu\sigma\omega)} \qquad (2.37)$$

δ is the classical skin depth for electromagnetic radiation in conducting media. In copper, taking $\mu = \mu_0$ and $\sigma = 6 \times 10^7$ S/m, the skin depth works out to 9 mm at 50 Hz and 65 μm at 1 MHz. Because electromagnetic waves do not penetrate conductors fully, AC transmission lines are effectively limited to diameters of a few times the skin depth. The problem can be reduced or eliminated by using polyphase or DC transmission.

Induction heating is a convenient way of raising the temperature in a body directly, without using an external heat source. An alternating electromagnetic field, produced usually by a water-cooled copper coil, induces currents in the workpiece (known as eddy currents). These currents dissipate energy by Joule heating and the temperature of the body rises from this internal heat. The skin depth is an important parameter in induction heating applications, since if it is too large compared to the size of the workpiece there is little absorption of power, and if it is too small the heating is confined to the surface of the specimen. With suitably designed coils for coupling the load to the source, it is possible to develop very large power densities in the desired part of the workpiece. In making skin depth calculations for induction heating purposes it is worth bearing in mind that σ is strongly dependent on temperature; and in the case of ferromagnetic samples, remember that μ is large below the Curie point, but not above it! (See problem 2.6.)

2.6 The Plasma Frequency

The plasma frequency corresponds to the frequency at which density waves in the free electron cloud charge (the plasma) oscillate. Let us see how these density waves arise. When the frequency of electromagnetic radiation is comparable to $1/\tau$, that is, $\omega\tau \approx 1$, then the equation of motion for the electron must be written

$$m(dv/dt + v/\tau) = - e\mathscr{E} \tag{2.38}$$

The mv/τ term is due to collisions and is of the same form as velocity damping of a mass on a spring. Supposing that v is of the form $v_0\exp(-j\omega t)$ and that $\mathscr{E}$ takes the same form as before, then (2.38) can be written

$$m(-j\omega t + 1/\tau)v_y = - e\mathscr{E}_y \tag{2.39}$$

for an electric field with only a y-component. And as $J_y = - nev_y$, substituting for v_y from (2.39) leads to

$$\begin{aligned}
J_y &= \frac{ne^2\mathscr{E}_y}{m\left[- j\omega + (1/\tau)\right]}\\
&= [ne^2\tau/m][\mathscr{E}_y(1 - j\omega\tau)]\\
&= \sigma\mathscr{E}_y/(1 - j\omega\tau) \tag{2.40}
\end{aligned}$$

since $\sigma = ne^2\tau/m$. Equation (2.4) is like $J_y = \sigma\mathscr{E}_y$, in which σ has been replaced by $\sigma/(1 - j\omega t)$, a complex conductivity. We can solve Maxwell's equations as before, when the skin depth was found, but we shall replace σ by its complex equivalent, so that (2.36) becomes

$$k^2 = \mu\epsilon\omega^2 + j\mu\sigma\omega/(1 - j\omega\tau) \tag{2.41}$$

If $\omega\tau \ll 1$, (2.41) reduces to (2.36), but if $\omega\tau \gg 1$ (that is, at 'high' frequencies) it becomes

$$\begin{aligned}
k^2 &= \mu\epsilon\omega^2 - \mu\sigma/\tau\\
&= \mu\epsilon\omega^2[1 - \sigma/\epsilon\tau\omega^2] \tag{2.42}
\end{aligned}$$

Writing ω^2_p for $\sigma/\epsilon\tau$, (2.42) is

$$\begin{aligned}
k^2 &= \mu\epsilon\omega^2(1 - \omega^2_p/\omega^2)\\
&= (\omega/v_p)^2(1 - \omega^2_p/\omega^2) \tag{2.43}
\end{aligned}$$

$\mu\epsilon$ has been replaced by $1/v_p$ where v_p is the phase velocity (ω/k) in the medium. Equation (2.43) is another dispersion relation, from which it may be seen that, if $\omega > \omega_p$, then k is real and the wave is unattenuated; above the plasma frequency, conductors are transparent to electromagnetic radiation. Below the plasma frequency, if $\omega\tau > 1$, the radiation is not absorbed, even though k is complex, but is totally reflected; absorption can only occur when $\omega\tau < 1$, or thereabouts – there has to be time for the collision mechanism to operate, if you like.

The transparency of conductors to electromagnetic radiation is contrary to common experience, but the plasma frequency is very high. Using

$$\omega^2_p \equiv \sigma/\epsilon\tau \tag{2.44}$$

we can readily find ω_p for copper. We have already found τ to be 2×10^{-14} s, ϵ is just ϵ_0 and $\sigma = 6 \times 10^7$ S/m, so (2.44) gives ω_p as 2×10^{16} rad/s, corresponding to a wavelength of 100 nm, which is in the ultra-violet. Metals must be transparent to ultra-violet radiation. When σ in (2.44) is replaced by $ne^2\tau/m$ we obtain

$$\omega_p = \sqrt{(ne^2/m\epsilon)} \tag{2.45}$$

Thus in most metals (but not semiconductors, and not all metals), where $\epsilon = \epsilon_0$, $m = 9.1 \times 10^{-31}$ kg and $e = 1.6 \times 10^{-19}$ C, we can write

$$\omega_p = 56\sqrt{n}$$

The plasma frequency depends only on the square root of the free electron concentration. The alkali metals are rather low in density and have only one free electron/atom, so we might expect to find the lowest plasma frequency and longest plasma wavelength in caesium, which may just be transparent to blue light; but it turns out not to be so by a narrow margin.

Table 2.6 shows calculated and experimental values for the plasma wavelength in some metals. λ_p is sometimes called the plasma cut-off wavelength, because at wavelengths longer than λ_p the plasma cuts off the transmission of electromagnetic radiation. The plasma wavelength has been calculated from equation (2.45) assuming that the number of free electrons is equal to the usual valency of the metal. The experimental values were obtained by looking for the ultra-violet cut-off wavelength or by measuring the energy lost from beams of electrons fired through thin metallic foils. The agreement between the calculated values of

Table 2.6 Plasma Wavelengths for some Metals

Metal	λ_p^C	λ_p^E
Lithium	155	174[*], 155[†]
Sodium	210	215[*], 210[†]
Potassium	284	324[*], 315[†]
Rubidium	314	340[†]
Caesium	360	440[†]
Magnesium	144	117[*]
Aluminium	79	81[*]

C = Calculated values of plasma wavelength.
E = Experimental values of plasma wavelength.
[*] = Value obtained from electron beams' energy-
      loss spectra.
[†] = Value obtained from ultra-violet cut-off.
All wavelengths in nm.

plasma wavelength and the experimental data from two different types of measurement is remarkably good.

2.6.1 An Example

The plasma cut-off for sodium is at 210 nm, its conductivity is 2.13×10^7 S/m and its Hall constant is -2.5×10^{-10} m³/C. Let us find the number of free electrons/atom, the mobility and the relaxation (collision) time. The Hall constant is $-1/ne$, giving n at once as 2.5×10^{28}/m³. From the Periodic Table we can see that sodium is bcc with lattice parameter 4.29 Å, so there are 2 atoms per unit cell, which has a volume of $(4.29 \times 10^{-10})^3$ m³, producing an atomic density of 2.53×10^{28}/m³, so there is just one free electron per atom.

But $\quad \sigma \quad = ne\mu$

so that $\quad \mu \quad = \sigma/ne = \sigma R_H$

$$= 2.3 \times 10^7 \times 2.5 \times 10^{-10} = 5.33 \times 10^{-3} \text{ m}^2/\text{V/s}$$

Since $\quad \omega_p^2 = \sigma/\epsilon\tau$

then $\quad \tau \quad = \sigma/\tau\omega^2$

$$= \sigma\lambda_p^2\epsilon/(2\pi c)^2$$

$$= 3 \times 10^{-14} \text{ s}$$

This is only true if $\omega_p\tau \gg 1$, so we must check this. ω_p works out to 9×10^{15} rad/s, so $\omega_p\tau$ is $9 \times 10^{15} \times 3 \times 10^{-14}$, which is 270, $\gg 1$ – the condition is met, the validity of the result is unquestioned.

2.6.2 Plasma Oscillations: Plasmons

We can calculate the plasma frequency from the consideration of the equations of motion for the electron gas. If the centre of charge of the free electron gas is displaced from the cations of the lattice by a small amount, ξ, then a field is set up whose magnitude may be calculated by Poisson's equation:

$$\epsilon\nabla.\mathscr{E} = q_v = -ne$$

q_v is the electric charge/m³, whch is just $-ne$. Integrating Poisson's equation gives

$$\epsilon\mathscr{E} = -ne\xi \tag{2.46}$$

This field exerts a force of $ne\mathscr{E}$ on the displaced electrons, so using $F = ma$, we have

$$ne\mathscr{E} = nmd^2\xi/dt^2 \tag{2.47}$$

The field is found from (2.46) and substituted into (2.47), producing

$$- ne^2\xi/\epsilon = md^2\xi/dt^2$$

We have simple harmonic motion, $d^2\xi/dt^2 = - \omega^2\xi$, in which the angular frequency is $\sqrt{(ne^2/m\epsilon)}$ – the plasma frequency. These oscillations in the free electron gas take the form of longitudinal density waves, like sound waves in a fluid, that can only propagate at a single frequency, the plasma frequency. A quantum of the longitudinal oscillation is known as a plasmon, which may be excited in a thin film of metal by firing an energetic beam of electrons at it. The beam loses energy to the plasmons, more or less in multiples of hf_p, where h is Planck's constant and f_p is the plasma frequency in Hz. The plasma frequency can be measured thus – see problem 2.4.

In silicon, the incorporation of an atom of phosphorus, arsenic or antimony into the lattice causes the liberation of an electron, which is available for conduction as in a metal. The spatial density of such electrons is often quite low by comparison with metals, so the plasma frequency is much lower and the plasma wavelength is longer – in the infra-red region. There is a minimum in reflectivity associated with the plasma frequency which is used as a standard test (described in ASTM[*] test method F398) for determining charge carrier concentration. See problem 2.5.

2.7 Failures of Classical Free Electron Theory

Though the free electron theory of conduction has taken us a long way – though not without hiccups, such as the disconcerting reversal in the sign of the Hall constant for certain metals – it has two major failings, which we shall examine next.

2.7.1 The Specific Heat of Metals

The classical theory of gases shows that the specific heat of a monatomic gas is $1\frac{1}{2}nk_BT$, or $1\frac{1}{2}R$ per mole, where R is the gas constant. Free electrons in metals appear to act as a monatomic ideal gas; thus we derived the Wiedemann–Franz law, which can only be explained if the electron gas acts like a gaseous atom in conducting heat, as well as electricity. Therefore in contemplating the specific heat of a metal at room temperature, we must conclude that it will be the same as an insulator's – are there not atoms in three-dimensional array capable of vibration like those of an insulator? – but with an additional contribution of $1\frac{1}{2}nk_B$ from the free electron gas. It

---

[*] The American Society for Testing and Materials, a body which has done a great deal to raise standards of production and testing in many industries, especially the electronic industry.

seems, then, that the Dulong and Petit specific heat for an insulator of $3R$ J/mole/K should be raised to $(3 + 1\frac{1}{2})R$ J/mole/K for a metal: it isn't – metals have almost exactly the same specific heats as insulators at room temperature or a bit above. Although electrons in metals appear to move so as to transport both heat and charge, they do not appear to absorb any energy when the temperature is raised, since they do not appear to contribute anything to the specific heat. Classical theory cannot resolve this paradox.

2.7.2 The Dependence of Electrical Conductivity on Temperature

Experimentally one finds that the dependence of electrical resistivity on temperature is, for pure metals, proportional to the absolute temperature, except at low temperatures, where it is usually taken as being proportional to T^5. The experimental data are very well fitted by the Grueneisen formula:

$$\rho(T)/\rho(\Theta) = 4.226G(\Theta/T) \tag{2.48}$$

The numerical constant must be $1/G(1)$, Θ is the Debye temperature of the metal, while $G(a)$ is the Grueneisen function, defined by

$$G(a) = a^{-5} \int_0^a \frac{b^5 db}{[\exp(b) - 1][1 - \exp(-b)]} \tag{2.49}$$

A plot of (2.48) for platinum is compared to experimental data in figure 2.6, taking Θ to be 230 K. On the scale of the drawing, no difference is discernible.

Conduction electrons are scattered by the thermal vibrations of the lattice, when no other imperfections are present, which cause a displacement, ξ, of atoms from their mean positions:

$$\xi = \xi_0 \sin\omega t$$

ω is the angular frequency of the vibrations, which have maximum amplitude, ξ_0. The force on the atom is then

$$F = md^2\xi/dt^2 = m\omega^2\xi_0\sin\omega t = m\omega^2\xi$$

The work done during the displacement is

$$W = \int_0^{\xi_0} Fd\xi = \tfrac{1}{2}m\omega^2\xi_0^2$$

This may be equated to the thermal energy, $\tfrac{1}{2}k_BT$. ω is determined by the stiffness of the bonding between atoms, which will be roughly the same for most solids from 0 to 300 K, so if

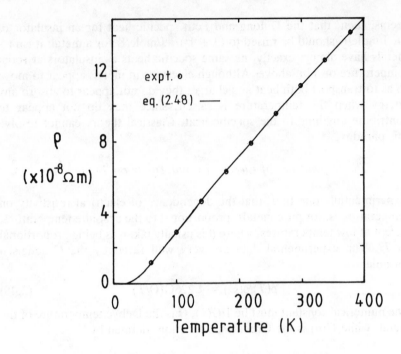

Figure 2.6 *The Grueneisen formula compared with experiment*

$$\tfrac{1}{2}m\omega\xi_0^2 = \tfrac{1}{2}k_B T$$

then ξ_0^2 is proportional to T, which means that the scattering cross-section of the lattice vibrations (proportional to ξ_0^2, probably) is proportional to the absolute temperature. The resistivity will be proportional to the scattering cross-section and thus to absolute temperature also, as observed, except at low temperatures, where there it goes as T^5. Only quantization of thermal vibrations can explain this low temperature behaviour.

Obviously, these are not the only failings of the classical free electron theory: the electrical properties of semiconductors have hardly been considered, for little progress is possible without jettisoning classical ideas about the behaviour of electrons. In subsequent chapters we shall retain the free electron theory, but the statistical behaviour of ensembles of electrons and lattice vibrations will not be classical. In this way the chief difficulties can be resolved.

Problems

2.1. A 100 W light bulb has a filament of tungsten wire 0.4 m long operating at 2850 K from a 240 V supply. If $\rho_{293} = 5.65 \times 10^{-8} \, \Omega\text{m}$

and the TCR is + 0.0055 over the temperature range 293–3000 K, what is the drift velocity of the electrons in the filament if each tungsten atom contributes one free electron? What is the ratio v_{th}/v_d at 2850 K?

[Ans. 7 mm/s; 5 × 10⁷]

2.2.   Figure P2.2 shows the phase diagram at the Cu-rich end of the Cu–Cr binary system. An alloy containing 1 per cent Cr and 99 per cent Cu is cooled slowly from 1100°C to 1071°C. Find the percentage of liquid at the latter temperature.

The alloy is then quenched to room temperature. Assuming that the eutectic liquid turns to α + Cr below 1070°C, what will be the percentage of α at room temperature?

It is found that 0.1 per cent of Cr in solid solution in Cu produces a 25 per cent increase in resistivity at room temperature. What will be the resistivity of the quenched alloy at room temperature if pure copper has a resistivity of 1.72 × 10⁻⁸ Ωm at 293 K? If the alloy is annealed for a long time at 600°C, what will then be its resistivity at room temperature and at 4.2 K?

Calculate the TCRs of both annealed and quenched alloys at room temperature if that of pure copper is + 0.004/K. What will be the effects of these heat treatments on the hardnesses of the alloys? Why?

[Ans. 43 per cent; 99.7 per cent; 47 nΩm; 3 nΩm; 0.0034/K and 0.00145/K]

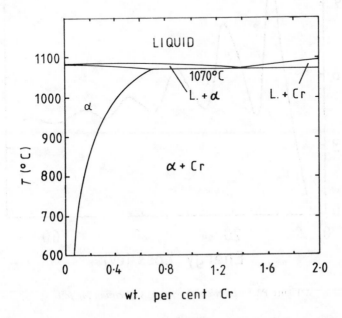

Figure P2.2   *The Cu–Cr phase diagram at the Cu-rich end*

2.3. Vacancies in platinum cause a residual resistivity of 4 pΩm for each ppm. At 1500°C the equilibrium concentration of vacancies is 0.2 per cent. If the energy of vacancy formation is 1 eV, what will be the resistivity ratio for pure platinum when it is annealed at 1000°C and then quenched to room temperature?

If the highest attainable resistivity ratio is 2000 and is due solely to vacancies, what is the maximum temperature at which platinum may be used as a resistance thermometer?

[ρ = 0.105 $\mu\Omega$m for pure platinum]

[Ans. 170; 730°C]

2.4. Figure P2.4 shows the relative absorption as a function of energy loss for a 2 keV beam of electrons which is reflected from a film of magnesium. The loss in energy is due to plasmon formation in the metal, so that $E_{loss} = nhf_p$ (n = 1,2,3...). Use all the absorption peaks to calculate the average value of hf_p and hence find the number of free electrons/atom in magnesium.

[Ans. Nearly 2]

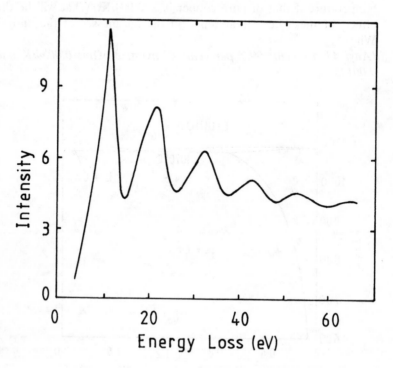

Figure P2.4   *Electronic energy-loss spectra in Mg foil*

2.5. To what depth will radio waves penetrate sea water, if its conductivity is 5 S/m? What significance does this have for communications?
[*Ans. About a metre*]

2.6. Industrial induction-heating processes often use mains frequency for convenience. What is the skin depth for iron at 300 K, 1000 K and 1200 K using mains-frequency induction heating? What sort of iron or steel artefacts might be so heated? [Assume $\mu_r = 500$ for iron below its Curie temperature of 1043 K. Use table 2.1 for the resistivity.]
[*Ans. 1 mm; 2 mm; 52 mm*]

2.7. A sample of phosphorus-doped silicon has a plasma resonance minimum in the infra-red region of 10 μm. What is the concentration of free electrons in the sample if $\epsilon_r = 12$ in silicon and equation (2.44) applies? If the resistivity is 10 $\mu\Omega$m, what is the drift velocity? [Remember $\epsilon = \epsilon_r\epsilon_0$. The effective mass of the electrons should be taken into account also, but this has not been discussed yet, and in any case the calculation is only approximate.]
[*Ans. 1.34 $\times$ 10^{26}/m^3; 4.7 mm/s*]

2.8. Electromagnetic waves are attenuated in coaxial cables at high enough frequencies because of absorption by the dielectric layer. A measure of the absorption is tanδ, which is 0.002 for polyethylene at 100 MHz. If δ is the same as 2ϕ in the equation for the complex wave vector, $k = K\exp(j\phi)$, use equation (2.36) to find the cable loss in dB/m at 100 MHz.
If the attenuation is 0.5 dB/m at 1 GHz, find tanδ at this frequency. [Take $\epsilon = 2.3\epsilon_0$ and $\mu = \mu_0$ for polyethylene. Note that the loss in dB is $-$ 8.686ln($\mathscr{E}/\mathscr{E}_0$).]
[*Ans. 0.028 dB/m; 0.036*]

2.9. Using equations (2.48) and (2.49) show that, for any metal, the TCR at the characteristic temperature, Θ, is $+ 1.11/\Theta$ per K. Θ for platinum is 230 K. If a piece of platinum wire has a resistance of 100 Ω at 230 K, what will be its resistance at 273.2 K?
[*Ans. 120.85 Ω. The experimental value is 120.95 Ω*]

2.10. The ionosphere reflects radio waves below a frequency of about 200 MHz. If this is the plasma cut-off, what is the electron density in the ionosphere? If each atom in the ionosphere contributes one electron to the plasma, what is the gas pressure in it, assuming its temperature is 300 K?
[*Ans. 5 $\times$ 10^{14}/m^3; 1.6 $\times$ 10^{-8} Torr*]

2.11. Show that the effective resistance of a conductor of circular cross-section is given by

$$R = R_0\rho^2[\exp(- 2\rho) + 2\rho - 1]/8[\exp(- \rho) + \rho - 1]^2$$

where R_0 is the DC resistance, ρ ($\equiv r/\delta$) is the reduced radius, r is the radius and δ the skin depth.

[The current density at a radius, a, from the centre of the cable will be $J(a) = J_s \exp[-(r-a)/\delta]$, where J_s is the current density at the surface. From this, find the total current in the wire in terms of J_s. Then find the power consumed per unit length of cable in a thin shell at radius, a. Integrate to find the total power/m and compare with the power lost with the same current in the wire and uniform current density.]

[For $r = \delta$, or $\rho = 1$, we find $R = 1.05R_0$]

3

The Quantum Theory of Electrons in Solids

In the previous chapter many of the electrical properties of solids were explicable in terms of a model in which the electron was merely a small, negatively charged particle obeying the Newton's laws of motion and Maxwell's equations. There were, however, serious difficulties: principally, the very small electronic specific heat and the wrong sign of the Hall coefficient in some metals and semiconductors. In the latter case, the device of positive charge carriers, which did not appear to exist in any other context, was an unsatisfactory *ad hoc* assumption. We can see now that Drude's free electron theory came at the very end of the classical era in physics – in fact it was published at the same time as Planck's quantum theory. However, though progress with these new theoretical concepts was rapid, it was not until 1931, when A. H. Wilson published two papers in the *Proceedings of the Royal Society* (entitled 'The Theory of Electronic Semi-conductors') that the theoretical basis of the electrical properties of solids was established.

Essentially, the new theoretical approach took away the certainty of the classical, particle theory and replaced it with an electron which behaved sometimes like a wave, sometimes like a particle, and whose momentum and position could only be described in terms of probabilities.

3.1 Schroedinger's Equation

A particle in quantum mechanics can be associated with a wave function, ψ (generally complex), such that the probability of finding the particle in a volume, dV, is given by $|\psi|^2 \, dV$. Schroedinger's equation enables us to find ψ, in principle, for any particle whose energy – kinetic and potential – is known:

$$\hbar^2 \nabla^2 \psi + 2m(E - U)\psi = 0 \tag{3.1}$$

61

Here, $\hbar$ is $h/2\pi$, where h is Planck's constant, m is the mass of the particle, E is the total energy of the particle and U is its potential energy. The use of wave functions to describe the behaviour of particles leads to results that are quite contrary to classical expectations, as we shall see now.

3.2   The Particle in a Potential Well

Consider a one-dimensional potential well as in figure 3.1, in which the potential energy of the particle is zero between $x = 0$ and $x = l$, and U_0 everywhere else, with $U_0 > E$. The Schroedinger equation then takes on two forms depending on which part of space is being described:

$$\hbar \partial^2\psi/\partial x^2 + 2m(E - U_0)\psi = 0 \qquad (3.2a)$$

outside the well, and

$$\hbar^2 \partial^2\psi/\partial x^2 + 2mE\psi = 0 \qquad (3.2b)$$

inside the well. The solution of (3.2a) is of the form

$$\psi = A\exp(k_1 x) + B\exp(-k_1 x)$$

where A and B are constants and k_1 is $\sqrt{[2m(U_0 - E)/\hbar^2]}$. The solution of (3.2b) is

$$\psi = C\sin(kx) + D\cos(kx)$$

where C and D are constants and k is $\sqrt{(2mE/\hbar^2)}$. Now the constants must be found by looking at the physical constraints – the boundary conditions. Firstly, in order for the particle to be somewhere in the well, the solution to (3.2a), which gives the probability of finding the particle outside the well, must go to zero as $x \rightarrow \pm \infty$, which means that when $x \leqslant 0$, $B = 0$; and when $x \geqslant l$, $A = 0$. The non-zero value of this solution means that there is

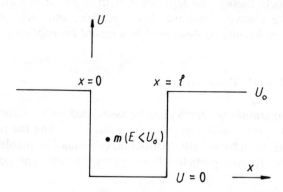

Figure 3.1   *A particle in a one-dimensional potential well*

a finite chance that the particle is outside the well even though its total energy, E, is less than the height of the potential barrier, U_0.

To simplify the solution, consider the case where $U_0 \to \infty$, so that $k_1 \to \infty$ and $\psi \to 0$ at $x = 0$ and $x = l$. But the solution for ψ inside the well must agree with this condition, that is

$$\psi = C\sin(kx) + D\cos(kx) = 0$$

when $x = 0$ and $x = l$. Substituting for $x = 0$ gives $D = 0$, so that the solution is just $\psi = C\sin(kx)$ for $0 \leq x \leq l$. Substituting for $x = l$ gives

$$C\sin(kl) = 0$$

which can only be true if $kl = n\pi$ ($n = 0, 1, 2 \ldots$), as the solution $C = 0$ means there is no wave function and no particle. But $k = \sqrt{(2mE/\hbar^2)}$, giving

$$k^2 l^2 = 2mEl^2/\hbar^2 = n^2\pi^2$$

or

$$E_n = n^2\pi^2\hbar^2/2ml^2 \tag{3.3}$$

where n is the quantum number of the energy state, E_n, which is the energy of the particle in the nth energy state. Equation (3.3) gives the energy *eigenvalues* of the particle: no others are permitted. The particle has a wavelength which can be found from the *de Broglie relation*

$$\lambda = h/p$$

where p is its momentum, mv. But in the potential well $U = 0$, so all the particle energy is kinetic energy

$$E = \tfrac{1}{2}mv^2 = p^2/2m$$
$$= h^2/2m\lambda^2$$

Thus

$$\lambda = h/\sqrt{(2mE)} \tag{3.4}$$

The wavelength of a particle, even as small as an electron, is very short. For instance, consider an electron which has kinetic energy of 1 eV or 1.6×10^{-19} J. Substituting for h ($= 6.626 \times 10^{-34}$ Js), m ($= 9.1 \times 10^{-31}$ kg) and E in (3.4) gives $\lambda = 1.22$ nm. An electron microscope has a resolving power far in excess of an optical microscope's for this reason.

3.3 The Pauli Exclusion Principle

One can consider a metallic crystal to be a three-dimensional infinite potential well whose boundaries are the crystal's surface and to which the free electrons are confined. The electrons may occupy a set of energy states like those of equation (3.3), but only two electrons are allowed in a given state, because of the *Pauli Exclusion Principle*. This states that each electron is characterized by a unique set of quantum numbers, including spin, so that the energy level with $n = 2$, say, can only contain two electrons: one spin up and one spin down.

To make matters simple, let us consider a one-dimensional crystal, comprising a line of length 10 mm, with atoms of 2 Å diameter, so that there are 5×10^7 atoms in the line, and suppose that each contributes one free electron. Since two electrons can occupy each level, 2.5×10^7 energy states are filled, and the maximum energy calculated from (3.3) is

$$E_{max} = \frac{(2.5 \times 10^7)^2 \pi^2 \hbar^2}{2 \times 9.1 \times 10^{-31} \times (10^{-2})^2}$$

$$= 3.8 \times 10^{-19} \text{ J} = 2.4 \text{ eV}$$

Equation (3.3) shows that E_{max} is proportional to $(n/l)^2$, that is, it depends only on the linear density of free electrons. At room temperature $E_{max} \approx 100 \, k_B T$. The energy levels are also very closely spaced, as may be seen by differentiating (3.3):

$$dE/dn = 2n\pi^2 \hbar^2 / 2ml^2 = 2E_n/n$$

so that

$$\Delta E = 2E_n \Delta n/n$$

and the maximum gap between adjacent energy levels ($\Delta n = 1$) when $n = 2.5 \times 10^7$ is $8 \times 10^{-8} E_{max}$, or about 0.2 μeV.

3.4 The Fermi Energy

The *fermi energy*, E_f, is the maximum energy that an electron may have in a crystal at 0 K. We have just made a rough sort of estimate for E_f which can be improved upon by modifying equation (3.3) to fit a three-dimensional potential well:

$$E = \frac{p^2}{2m} = \frac{\pi^2 \hbar^2}{2ml^2} (n_x^2 + n_y^2 + n_z^2)$$

giving

$$(n_x^2 + n_y^2 + n_z^2) = \frac{2ml^2}{\pi^2 \hbar^2} E = R^2 \tag{3.5}$$

n_x, n_y and n_z are quantum numbers associated with the x-, y- and z-axes of the crystal. Each state has three quantum numbers to describe it, and as electrons are added to the crystal, the states fill up in turn with two electrons each, like an expanding sphere of radius, R, in n-space, as shown in figure 3.2. For a macroscopic crystal, the magnitude of n is very large and n-space is almost a continuum. Since only positive values of n are allowed, the volume occupied in n-space is one-eighth of a sphere of radius R, or $\frac{1}{6}\pi R^3$. If there are N electrons in the crystal, these must fill $N/2$ states, so that

$$\frac{1}{6}\pi R_{max}^{3} = \frac{1}{2}N$$

$$R_{max} = (3N/\pi)^{\frac{1}{3}}$$

where R_{max} is the maximum radius reached in n-space. Substituting for R_{max} in (3.5) yields

$$\frac{2ml^2}{\pi^2\hbar^2}E_{max} = (3N/\pi)^{\frac{2}{3}}$$

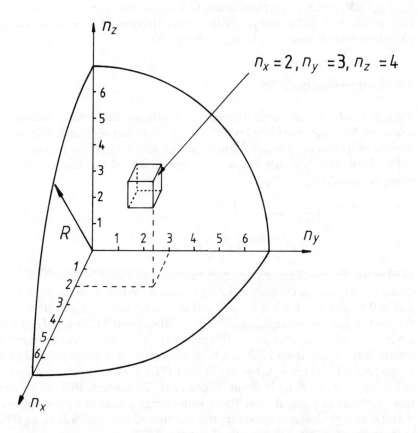

Figure 3.2 *Filled states in n-space*

But E_{max} is E_f by definition, so

$$E_f = \frac{\pi^2\hbar^2}{2ml^2} (3N/\pi)^{\frac{2}{3}} = \frac{\pi^2\hbar^2}{2m} (3N/\pi l^3)^{\frac{2}{3}}$$

Now N/l^3 is n_e, the number of electrons/ unit volume, giving

$$E_f = \frac{\pi^2\hbar^2}{2m} (3n_e/\pi)^{\frac{2}{3}} \qquad (3.6)$$

$$= 3.65 \times 10^{-19} n_e^{\frac{2}{3}} \text{ eV} \qquad (3.7)$$

The fermi energy thus depends solely on the concentration of electrons in the solid, and can now be calculated more accurately. Silver, for example, has the fcc structure with $a = 4.1$ Å. There are 4 atoms/unit cell, each contributing one free electron (as the valency of silver is one), so n_e is $4/(4.1 \times 10^{-10})^3$, or 5.8×10^{28}/m^3, and substitution in (3.7) gives $E_f = 5.5$ eV, which is a typical value. In a metal, the fermi energy will not vary much with temperature as the concentration of free electrons is virtually constant from 0 K to the melting point.

3.5 Fermi–Dirac Statistics

Though we have calculated the energy levels available to electrons in solids, we have not considered whether they are in fact occupied. Electrons are one of a class of particles (protons are another) which are said to obey *Fermi–Dirac statistics*, that is to say, the probability of an energy state, E, being occupied is given by

$$F(E) = \frac{1}{1 + \exp\left[\dfrac{E - E_f}{k_B T}\right]} \qquad (3.8)$$

$F(E)$ is the Fermi–Dirac distribution function, and may be interpreted as giving the average occupancy of energy state, E. When $E = E_f$, $F(E) = \frac{1}{2}$, and at 0 K, $F(E) = 1$ when $E < E_f$ and is zero when $E > E_f$, as shown by the dotted line in figure 3.3(a). At its melting point (1234 K), $F(E)$ for silver looks like the solid line in figure 3.3(a). At room temperature $F(E)$ differs only slightly from $F(E)$ at 0 K, as may be seen in the expanded plot in figure 3.3(b). If $(E - E_f) = 3k_B T$, then $F(E)$ is about 5 per cent, and if $(E_f - E) = 3k_B T$, $F(E)$ is about 95 per cent. This means that, of all the free electrons in a metal, only those with energies near to E_f can acquire thermal energy; roughly speaking, this fraction of the total will be $k_B T/E_f$ so that in silver at 300 K only about 0.025/5.5, or 0.5 per cent, can contribute to the specific heat.

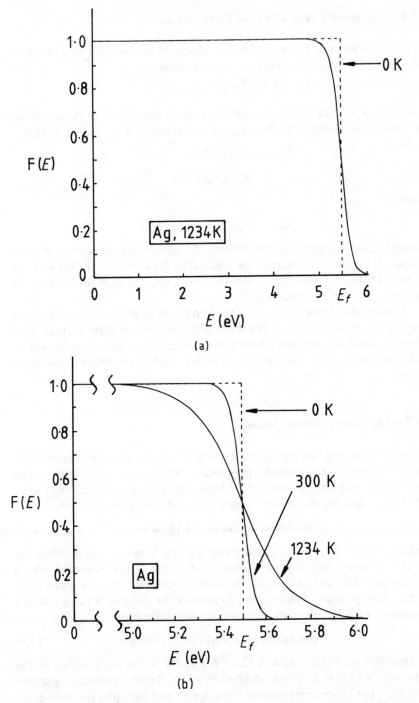

Figure 3.3 *(a) The fermi function for Ag; (b) the fermi function for Ag near E_f*

3.6 The Specific Heat of a Free Electron Gas

Let the number of thermally excited electrons/unit volume in a metal be N_{th}, then the thermal energy of these electrons is

$$E_e = N_{th} \times 1\tfrac{1}{2}k_B T$$

as there is $\tfrac{1}{2}k_B T$ of thermal energy for each of the electron's three degrees of translational freedom. But N_{th} is approximately $n_e \times k_B T/E_f$ so that

$$E_e \approx n_e \times (k_B T/E_f) \times 1\tfrac{1}{2}k_B T$$

$$\approx 1\tfrac{1}{2}n_e k_B^2 T^2/E_f$$

and so

$$C_e = \partial E_e/\partial T \approx 3 n_e k_B^2 T/E_f$$

where C_e is the electronic specific heat. (A more rigorous derivation of C_e makes the numerical constant $\tfrac{1}{2}\pi^2$ instead of 3.) For silver at 300 K, C_e is about 1 J/kg, compared with the measured specific heat of 234 J/kg, or about $\tfrac{1}{2}$ per cent, as expected.

Already these new ideas about the behaviour of electrons in solids have resolved a major difficulty of the classical theory, but there is quite a lot more to be done. So far we have not considered how the electrons will be affected by the atoms in the crystal lattice, and this problem is discussed next.

3.7 The Penney–Kronig Model

Electrons moving in a crystal are not really free, but must be influenced to some degree by the periodic potential of the atoms in the lattice. The problem is to formulate the Schroedinger equation for this and then solve it. In one dimension, Schroedinger's equation can be written as

$$\hbar^2 \partial^2 \psi/\partial x^2 + 2m[E - U(x)]\psi = 0$$

where $U(x) = U(x + a)$, a being the spacing between atoms along the x-axis. Penney and Kronig suggested that the lattice potential was a rectangular barrier of height U_0 and width δ, such that $U_0 \to \infty$ and $\delta \to 0$, while their product, $U_0\delta$, remained constant. In this case it turns out that solutions to Schroedinger's equation are only possible if

$$\cos(ka) = P\sin(k_1 a)/k_1 a + \cos(k_1 a) \qquad (3.9)$$

where $P = maU_0\delta/\hbar^2$, $k_1 = \sqrt{(2mE/\hbar^2)}$ and k is the wave vector of the electron, which is, in effect, defined by (3.9). To interpret this equation, we must put in some reasonable value for P, and then plot the right-hand side as a function of k_1a. Taking the atomic spacing, $a = 2$ Å, $\delta = 2$ Å and $U_0 = 7$ eV leads to $P \approx 4$, and the plot of the right-hand side of (3.9) looks as in figure 3.4.

Now $\cos(ka)$ can only take on values between -1 and $+1$, so for parts of the graph there are *forbidden gaps* – values of k_1 which are not allowed. At low values of k_1a the forbidden energy gaps are wide, but at higher values they become smaller and smaller. When $k_1a = \pm n\pi$, (3.9) becomes just $\cos(ka) = \cos(k_1a)$, and $k = \pm n\pi/a$. These values of k are points of discontinuity in the $E(k)$ curve for electrons in the crystal.

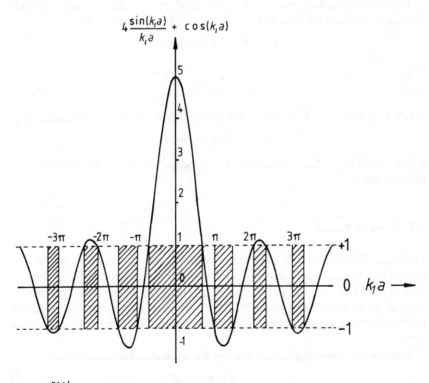

$\boxed{\!/\!/} =$ forbidden values

Figure 3.4 *The graph of equation (3.9), the Penney–Kronig solution*

Equation (3.9) can be shown to reduce to the correct physical result in two extreme cases:

(1) the lattice potential barrier, $U_0\delta$, is very large
(2) the lattice potential is very small.

In the first case, if $U_0\delta$ is very big, so then is P, so that (3.9) reduces to

$$P\sin(k_1a) = 0$$

Hence $k_1a = n\pi$ and, as k_1 is $\sqrt{(2mE/\hbar^2)}$, it produces

$$E_n = n^2\pi^2\hbar^2/2ma^2$$

which is the same as (3.3), the solution to the infinite potential well problem, with the size of the well being a, the distance between lattice sites. In other words, in an infinite lattice potential the electrons are bound to the atoms, just as expected.

The second limiting case corresponds to $P = 0$, so that equation (3.9) is reduced to $\cos(ka) = \cos(k_1a)$, or $k = k_1$, which means that

$$k = k_1 = \sqrt{(2mE/\hbar^2)}$$

or

$$E = \hbar^2k^2/2m \tag{3.10}$$

Since $k = 2\pi/\lambda = p/\hbar$ by the de Broglie relation, (3.10) can be reduced to

$$E = p^2/2m = \tfrac{1}{2}mv^2$$

which is the free electron energy, the expected result in the absence of a lattice potential.

3.8 Energy Bands

Equation (3.10) gives an $E(k)$ curve which is parabolic, as shown by the dotted line in figure 3.5, while an electron acted on by the lattice potential of a crystal has an $E(k)$ curve shown by the solid lines. Near the energy gaps at $ka = \pm n\pi$ the solid lines deviate markedly from the free electron parabola, but at some remove from the discontinuities, they are very much alike.

When $k = \pm n\pi/a$, (3.10) gives for the free electron energy

$$E = n^2\pi^2\hbar^2/2ma^2 \tag{3.11}$$

which is exactly the same as (3.3), the solution for the potential well problem, with l, the size of the well, replaced by a, the distance between atoms. If $a = 3$ Å, then with $m = 9.1 \times 10^{-31}$ kg, (3.11) can be written

$$E \approx 4n^2 \text{ eV}$$

Thus, beyond the first few discontinuities in the $E(k)$ curve, the electron's energy becomes very large and we are not usually concerned with them. The regions between the energy discontinuities are called *Brillouin zones*, labelled 1st B. Z., 2nd B. Z. etc. in figure 3.5(a). By folding higher zones

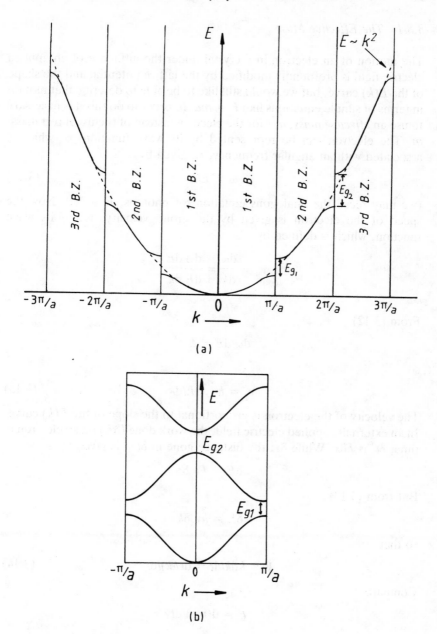

Figure 3.5 *(a) The E(k) curve for a nearly-free electron; (b) the reduced zone scheme*

into the first zone, the $E(k)$ diagram can be condensed into a smaller compass, known as the *reduced zone scheme*, as in figure 3.5(b).

3.8.1 The Effective Mass

The motion of an electron in a crystal under the influence of an applied electric field is profoundly modified by the lattice potential and the shape of the $E(k)$ curve, but we would still like to be able to describe that motion in terms of simple equations like $F = ma$. In order to do this it is necessary to use an *effective mass*, m^*, for the electron instead of the usual rest mass, m. The electron can be represented by its wave function, ψ, which is associated with an angular frequency, ω, given by

$$\omega = E/\hbar \tag{3.12}$$

(we may recall the analogous equation for photons, $E = \hbar\omega$). Now the speed of the electron is given by the group velocity, v_g, of its wave function, which is defined by

$$v_g = \frac{d\omega}{dk} = \frac{d\omega}{dE}\frac{dE}{dk}$$

From (3.12)

$$d\omega/dE = \hbar^{-1}$$

so

$$v_g = \hbar^{-1}dE/dk \tag{3.13}$$

The velocity of the electron is proportional to the slope of the $E(k)$ curve. In an externally applied electric field, the work done (δE) on an electron in time, δt, is $F\delta x$. While δx, the distance gone in δt s, is given by

$$\delta E = Fv_g\delta t$$

But from (3.13)

$$\delta E = \hbar v_g \delta k$$

so that

$$F = \hbar dk/dt = d(\hbar k)/dt \tag{3.14}$$

Compare

$$F = d(mv)/dt$$

Because of the formal similarity of these two equations $\hbar k$ is known as the *crystal momentum* of the electron. By differentiating (3.13) with respect to t, we have

$$dv_g/dt = \hbar^{-1}\partial^2 E/\partial k \partial t = \hbar^{-1}(\partial^2 E/\partial k^2)(dk/dt)$$

from which

$$\hbar dk/dt = \hbar^2(\partial^2 E/\partial k^2)^{-1} dv_g/dt \tag{3.15}$$

$$= F \qquad \text{from (3.14)}$$

Now

$$F = m^*a = m^* dv_g/dt \tag{3.16}$$

Comparing (3.15) and (3.16), we can see that the effective mass is given by

$$m^* = \hbar^2(\partial^2 E/\partial k^2)^{-1} \tag{3.17}$$

Since the curvature of the $E(k)$ graph is related to the quantity $(\partial^2 E/\partial k^2)^{-1}$, a high curvature corresponds to a large value of the effective mass.

It can be shown (see problem 3.1) that equation (3.17) gives $m^* = m$, for a free electron, as it should. However, inside a crystal, where the $E(k)$ curves are not parabolic, the effective mass takes on some surprising values. For example, consider the case where the $E(k)$ curve is as shown in figure 3.6, in which the lower curve can be expressed as

$$E = E_0 \cos(ka) - E_0$$

for small values of ka, so that

$$\partial^2 E/\partial k^2 = -a^2 E_0 \cos(ka) \approx -a^2 E_0$$

near $k = 0$, in which case the effective mass is $-\hbar^2/a^2 E_0$, a negative quantity as E_0 is positive. Rather than have a negatively charged particle of

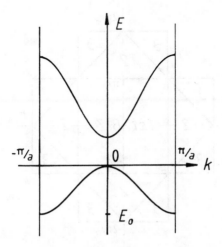

Figure 3.6 *An E(k) curve giving negative effective mass*

negative effective mass, one can take both the mass and charge to be positive, this new particle being known as a hole. Holes are vacant states into which electrons can move under, for example, the influence of an electric field. When a hole is filled by an electron moving along the $-x$ direction it is equivalent to a positive charge moving in the $+x$ direction. Typically, $E_0 \approx 1$ eV and $a \approx 2$ Å and then m^* works out to 1.7×10^{-30} kg. Effective masses of holes and electrons are expressed in terms of the electronic rest mass, m_0; here the effective mass of a hole at the top of the valence band near $k = 0$ is $1.9m_0$.

3.8.2 Brillouin Zones

In one dimension the points of energy discontinuity are, as we have seen, at $k = \pm n\pi/a$. In two dimensions, the equivalent expression, stated without proof, is

$$n_x k_x + n_y k_y = (n^2_x + n^2_y)\pi/a$$

which defines a set of straight lines bounding the Brillouin zones of a two-dimensional crystal, which is a simple square lattice. For the first zone, the bounding lines have $n_x = 0$ and $n_y = \pm 1$; and $n_x = \pm 1$, $n_y = 0$. The lines are then just $k_x = \pm\pi/a$ and $k_y = \pm \pi/a$, which defines the square first zone shown in figure 3.7. The second zone has $n_x = n_y = \pm 1$, so the zone boundaries are the lines $k_x + k_y = \pm 2\pi/a$ etc., which are shown also

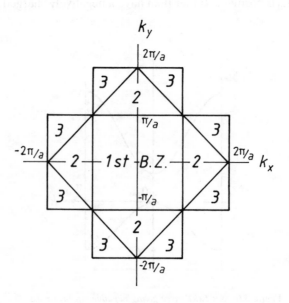

Figure 3.7 *Brillouin zones for a square lattice*

in figure 3.7, enclosing the whole of the first zone. The third zone is more complicated and its derivation is left as an exercise. The zones have equal areas.

In three dimensions the zone boundaries are defined by planes. For a simple cubic lattice the first zone boundary is a cube, whose cross-section is the first zone boundary of the square lattice. The higher zone boundaries also have cross-sections which are the boundaries of the Brillouin zones of the square lattice. Both bcc and fcc lattices have more complicated zone boundaries than the simple cubic structure.

3.8.3 The Fermi Surface

The *fermi surface* is the energy contour in k-space (also known as reciprocal-lattice space, momentum space etc.) for electrons with the maximum allowable energy, that is, the fermi energy, E_f. The fermi surface is where the action is; electrons with energies a bit below E_f cannot take part in conduction processes because there are no available states – only electrons which are at, or very near, the fermi surface can participate.

Electrons at the fermi surface have high energies and hence high velocities: of the order of 10^6 m/s (see problem 3.9). The relaxation time for collisions, τ, was found to be about 2×10^{-14} s in chapter 2, so that the mean free path λ $(= v_f\tau) \approx 2 \times 10^{-8}$ m or 20 nm, which is about ten times as big as the previous estimate. The process of conduction is now to be pictured as a displacement of all the electrons inside the fermi surface under the influence of an applied field, with only the most energetic electrons at the surface being scattered by collisions into lower energy states, as in figure 3.8. The drift velocity has the same magnitude as found in chapter 2, and from it can be found the displacement of the fermi sphere, Δk:

$$v_d = \hbar \Delta k / m^*$$

Suppose we add electrons to a two-dimensional square lattice, then at various stages in the filling of the Brillouin zones we can plot equal-energy contours, which might look as shown in figure 3.9. At small values of k the contours are circles (they would be spheres in three dimensions), as the $E(k)$ curve is quadratic for small k-values (that is, near the free electron value of energy, where the influence of the lattice potential is small). In this example, as the electron's energy increases, however, the fermi surface is distorted, and in particular is pulled out towards the zone boundary in certain crystallographic directions. In the <10> directions the allowed k-values are higher for a given energy than in the <11> directions, so that an electron in zone 2 in the <10> direction can have a lower energy than an electron in zone 1 in the <11> direction. The zones are said to overlap, and the $E(k)$ diagram would look as in figure 3.10(a).

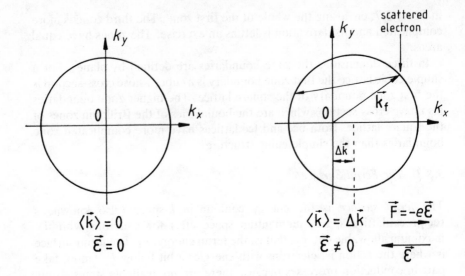

Figure 3.8 *The displacement of the free-electron fermi sphere by an electric field*

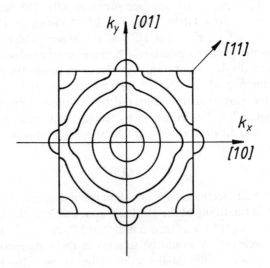

Figure 3.9 *The equal-energy contours of a square lattice*

It may happen that, though the fermi surface is distorted near the boundaries of a zone, there is no zone overlap, so that the $E(k)$ curve is as in figure 3.10(b). Even though the fermi surfaces in the two cases are almost identical, except at the zone boundary, the difference is crucial to

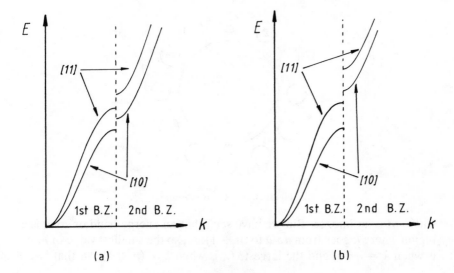

Figure 3.10   *(a) The E(k) curve for overlapping zones; (b) the E(k) curve with no zone overlap*

the electrical behaviour of the material, which will be more clearly understood when we find out how many energy states are available in each zone.

3.8.4   The Density of States

Consider a one-dimensional line, length l, of N atoms distance a apart, arranged in a ring as in figure 3.11. The wave functions at the beginning and end of the line must be the same, that is

$$\psi(x + l) = \psi(x) \tag{3.18}$$

Now the solution to the one-dimensional Schroedinger equation in a periodic potential takes the form

$$\psi(x) = \exp(jkx)u_k(x) \tag{3.19}$$

where $u_k(x)$ is such that $u_k(x + na) = u_k(x)$, where n is any integer. (u_k is known as a Bloch function after its inventor.) Since $l = Na$, substitution of (3.19) into (3.18) leads to

$$\exp(jk(x + l)) = \exp(jkx)$$

so that

$$kl = 2m\pi$$

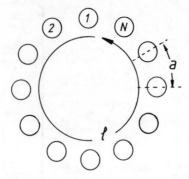

Figure 3.11   *Atoms in a ring*

where m is an integer. But we have seen that an energy band or zone is a region where k goes from $n\pi/a$ to $(n + 1)\pi/a$, so the smallest value of m, s, is when $k = n\pi/a$, and the largest, L, is when $k = (n + 1)\pi/a$, that is

$$s = k_{min}l/2\pi = nl/2a$$

$$L = k_{max}l/2\pi = (n + 1)l/2a$$

And

$$\Delta m = L - s = l/2a$$

This is the number of k-values in a band, except that m can take on both positive and negative values, and moreover, two electrons are allowed (spin up and spin down) for each value of k, so that there can be a maximum of $4\Delta m$, or $2l/a$, electrons in a band. Since $l/a = N$, the band is filled by $2N$ electrons, exactly twice the number of atoms in the ring.

In a three-dimensional crystal one can write

$$k_x = 2m_x\pi/l_x$$

with similar expressions for the y- and z-directions. The volume in k-space corresponding to a single level is $(2\pi)^3/l_xl_yl_z$, or $(2\pi)^3/V$, where V is the volume of the crystal. In k-space the volume of a sphere is $\frac{4}{3}\pi k^3$, so the number of states in k-space will be

$$N = 4\pi k^3/3 \div (2\pi)^3/V = Vk^3/6\pi^2$$

and the number of electrons, N_e, is $2N$ ($= Vk^3/3\pi^2$). For a free electron, $k = \sqrt{(2mE/\hbar^2)}$, giving

$$n_e = N_e/V = (1/3\pi^2)(2mE/\hbar^2)^{\frac{1}{2}}$$

$$= (1/3\pi^2)(2m/\hbar^2)^{\frac{1}{2}}E^{\frac{1}{2}} \qquad (3.20)$$

The number of states per unit energy is

$$D(E) = dn_e/dE = (1/2\pi^2)(2m/\hbar^2)^{\frac{1}{2}}E^{\frac{1}{2}} \qquad (3.21)$$

$D(E)$ is called the *density of states* and though it tells us how many electron states are available in a given energy range, it does not tell us whether these are occupied or not, which depends on the fermi function, $F(E)$. Hence the number of electrons per unit energy range, $N(E)$, will be

$$N(E) = D(E)F(E)$$

Of course, $F(E) \approx 1$ when $E < E_f$ for normal values of E_f and temperature, so $N(E) \approx D(E)$ when $E < E_f$ and $N(E) \approx 0$ when $E > E_f$. Figure 3.12(a) shows $N(E)$ against E for $T \approx 1000$ K. The density of states increases rather sharply at the zone boundary, and also at the places where zones overlap; figure 3.12(b) shows the general shape of the graph.

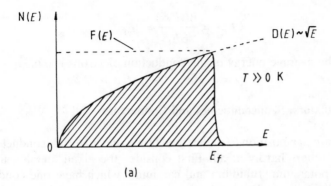

(a)

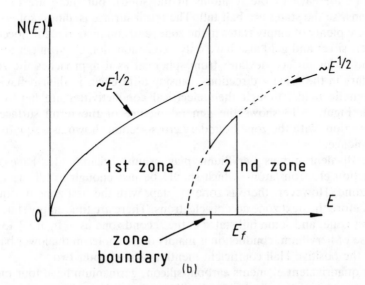

(b)

Figure 3.12   (a) *The density of states for a free electron; (b) the density of states when zones overlap*

Equation (3.21) allows us to calculate the average energy, $<E>$, of the conduction electrons, which is

$$<E> = \frac{\int_0^\infty N(E)E\,dE}{n_e} = \frac{\int_0^{E_f} D(E)F(E)E\,dE}{n_e}$$

$$= \frac{\int_0^{E_f} D(E)E\,dE}{n_e} = \frac{\beta \int_0^{E_f} E^{1\frac{1}{2}}dE}{\frac{2}{3}\beta E_f^{1\frac{1}{2}}} \tag{3.22}$$

This at 0 K, since $F(E) = 1$ from $E = 0$ to $E = E_f$ and is zero for $E > E_f$ while $\beta = (1/2\pi^2)(2m/\hbar^2)^{1\frac{1}{2}}$. n_e is found from (3.20), with $E = E_f$. Integrating (3.22) gives

$$<E> = \frac{\beta(\frac{2}{5}E_f^{2\frac{1}{2}})}{\frac{2}{3}\beta E_f^{1\frac{1}{2}}} = \frac{3}{5}E_f$$

so that the average energy of the conduction electrons is $0.6E_f$.

3.9   Insulators, Semiconductors and Conductors

Now we are in a position to understand how some materials conduct well at 0 K and others hardly at all. First consider the alkali metals – lithium, sodium, potassium, rubidium and caesium – which have one conduction electron per atom. The Brillouin zone containing the electrons has $2N$ states (2 for each of the N atoms in the solid), but there are only N electrons, so the states are half full. The fermi surface is almost spherical, there are plenty of empty states in the zone, and conduction readily occurs. Copper, silver and gold also have only one conduction electron per atom, but the fermi surface deviates from spherical as it approaches the zone boundary in the $<111>$ directions, causing differences in their behaviour in magnetic fields, though their electrical conductivities are hardly affected. Figure 3.13 shows the general picture of the fermi surface in cross-section, with the zone boundary cross-section shown as a square for convenience.

The divalent metals beryllium, magnesium, calcium etc. have two conduction electrons/atom, which should be just enough to fill the Brillouin zone. However, there is zone overlap, with the first zone not quite filled before the next zone accepts electrons. There are thus empty states in the first zone, and some full states in the second zone as in figure 3.13. In the case of beryllium, conduction is mainly due to holes in the lower band, hence the positive Hall coefficient mentioned in chapter two.

The quadrivalent elements carbon, silicon, germanium have four outer electrons, which are just sufficient to fill two zones. There is no overlap and

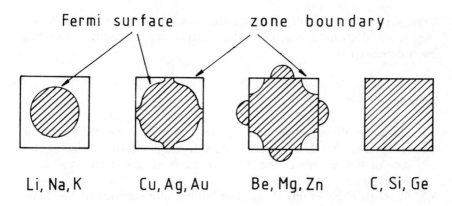

Figure 3.13 *The fermi surfaces in cross-section for various elements*

a band gap exists in all three cases, however its magnitude decreases very rapidly with increasing lattice parameter (all three have the diamond structure – we are ignoring the special case of graphite). Diamond has a band gap of about 5 eV, so there is virtually no possibility of thermal excitation of an electron across the gap, but silicon has a band gap of only about 1 eV, which is small enough to allow some electrons to be thermally excited across the gap and for conduction to occur: however, at room temperature, pure silicon is a poor conductor unless electrically active impurities are incorporated into the lattice (a process called *doping*). Germanium has a band gap of about 0.66 eV, and is accordingly more conducting than silicon at the same temperature. Diamond is therefore an insulator while silicon and germanium are semiconductors at temperatures around 300 K; yet it must be said that the difference in conducting behaviour is really only one of degree – at 0 K all three will be insulators.

The elements arsenic, antimony and bismuth are semimetals in which there is band overlap to a small degree, with the fermi level roughly in the middle of the 'negative gap', as in figure 3.14. The result of this is that the

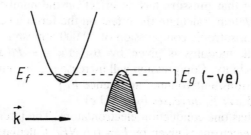

Figure 3.14 *The band structure of the semi-metals*

fermi surface is of small area and the number of states available for conduction electrons is much less than in true metals, resulting in much lower conductivity.

Problems

3.1. Show that the effective mass of a free electron is just the electronic rest mass, that is $m^* = m_0$. Draw a graph of m^* against k for $0 \leqslant ka \leqslant \pm \pi$, when the $E(k)$ curve is of the form $E = E_0 - E_0\cos(ka)$.

3.2. Show that the Fermi–Dirac distribution approximates to the Maxwell–Boltzmann distribution when $(E - E_f) \gg k_B T$. This is sometimes known as the Boltzmann tail, and is often a useful approximation. What is the relative error in this approximation where $E = E_f + 3k_B T$?
[Ans. 5 per cent]

3.3. Potassium has one conduction electron per atom. What will be its fermi energy? What energy corresponds to a probability of occupancy of 0.99 at 300 K? And to a probability of 0.01 and 300 K?
[Use the Periodic Table to find the structure and hence the number of atoms/unit volume.]
[Ans. 2.1 eV; 2.0 eV; 2.2 eV]

3.4. Aluminium has a fermi energy of 11.7 eV. How many conduction electrons/atom are there?
[Ans. 3, as might be expected]

3.5. When two metals are joined together, their fermi levels must be equal, so there is a transfer of electrons from the metal of higher fermi level to the one of lower until the contact potential is equal to the original difference in fermi level. Estimate the contact potential for copper and gold.
[Ans. 1.5 V]

3.6. What will be the fermi level in silver if it has one conduction electron/atom? The bulk elastic modulus for silver is about 110 GPa. Assuming that pressure has no effect on the number of conduction electrons/atom, calculate the effect on the fermi level of the application of an isotropic compression of 10 000 atmospheres.
[The bulk modulus is given by $B = 1/\kappa = - P/(\Delta V/V)$. κ is the compressibility. Compression will increase n_e, the number of conduction electrons/unit volume, and hence E_f.]
[Ans. 5.5 eV; E_f increases by 0.033 eV]

3.7. Sodium has one conduction electron/atom. The internal energy of N of these electrons is given by $E = 0.6 N E_f$. Calculate the compressibility of the conduction electrons in sodium.

[Use − $dE/dV = P$ and the compressibility, $\kappa = -V^{-1}(dV/dP)$. E_f is given by equation (3.6).]

[*Ans. 1.2×10^{-10} m^2/N. The measured value is 1.5×10^{-10} m^2/N*]

3.8. The thermal expansion coefficient is given by $\alpha = l_0^{-1}\Delta l/\Delta T$, where l_0 is the original length and Δl is the expansion for a rise in temperature of ΔT. If the fermi level changes only as a result of a change in the concentration of free electrons and the number of free electrons/atom is independent of temperature, show that $\Delta E_f/E_f \approx -2\alpha\Delta T$, if $\alpha\Delta T \ll 1$. If $\alpha = 19 \times 10^{-6}/K$ for silver, what will be the change in its fermi level when the temperature is raised from 0 K to its melting point (1234 K)?

[*Ans. − 0.26 eV*]

3.9. Find the group velocity of the electron wave-packet at the (spherical) fermi surface of lithium.

[Use equation (3.13). First calculate E_f and then use the free electron $E_f - k_f$ relation. v_g is v_f, the fermi velocity.]

Lithium has an electrical conductivity of 12 MS/m at 300 K. Estimate the relaxation time and hence the mean free path of the electrons which are scattered at the fermi surface.

[*Ans. 1.3×10^6 m/s; 9.2×10^{-15} s; 12 nm*]

4

Charge Carriers in Semiconductors

With the introduction of the concepts of the Pauli principle, the fermi level, energy bands and holes, we are now in a position to look in more detail at the behaviour of electrons and holes in semiconductors, which will lead to an understanding of the operation of devices – particularly diodes and transistors.

4.1 Intrinsic Conduction in Semiconductors

The previous chapter has shown us that a semiconductor is just an insulator which is not very good at its task, because electrons can be excited across the band gap at room temperature. Each electron that leaves the lower, or *valence* band, and moves into the upper, or *conduction* band, leaves a hole behind. This is an important fact: in an *intrinsic* semiconductor.

$$n_i = p_i$$

where n_i = number of conduction-band electrons/m^3 and p_i = number of valence-band holes/m^3. An intrinsic semiconductor is one in which electrical conduction is due to electrons and holes produced in this way. The question is: what are n_i and p_i? Classically, we would write

$$n_i = N_c \exp(- E_g/k_B T) \qquad \text{[INCORRECT!]}$$

where E_g is the band gap energy and N_c is related to the number of states in the conduction band. It is to be hoped that we now know better and write instead

$$n_i = \int N(E)F(E)\mathrm{d}E$$

84

with limits of integration going from the top of the valence band to infinity. The situation might look like figure 4.1, where for convenience we have made the not-unrealistic assumption that the bands are parabolic (that is, the free-electron approximation).

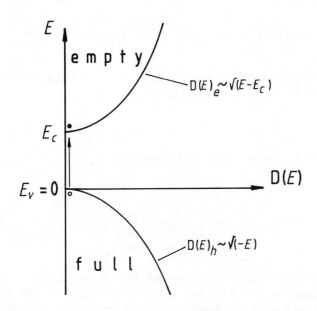

Figure 4.1   *D(E) against E near the band edges in a semiconductor*

For the electrons in the conduction band

$$D(E)_e = \beta_e \sqrt{(E - E_g)}$$

where β_e is $(2\pi^2)^{-1}(2m_e{}^*/\hbar^2)^{\frac{1}{2}}$ as defined in (3.21), $m_e{}^*$ is the electronic effective mass and we have taken the top of the valence band to be our energy zero. For holes we can write

$$D(E)_h = \beta_h \sqrt{(-E)}$$

where β_h is $(2\pi^2)^{-1}(2m_h{}^*/\hbar^2)^{\frac{1}{2}}$ and $m_h{}^*$ is the effective mass for holes. (E is negative in the valence band.) It turns out that in the conduction band $(E - E_f) \gg k_B T$, in which case

$$F(E) \approx \exp(-[E - E_f]/k_B T)$$

$$n_i = D(E)_e F(E)$$

$$= \beta_e \sqrt{(E - E_g)} \exp(-[E - E_f]/k_B T)$$

Making the substitution $x = (E - E_g)/k_B T$, so that $dx = dE/k_B T$, and writing $(E_g - E) + (E - E_f)$ for $(E - E_f)$ leads to

$$n_i = \int_{E_g}^{\infty} N(E)_e dE$$

$$= \beta_e (k_B T)^{1\frac{1}{2}} \exp(- [E_g - E_f]/k_B T) \int_0^{\infty} x^{\frac{1}{2}} \exp(- x) dx$$

The limits of integration for electrons are from the bottom of the valence band (where $E = E_g$, so $x = 0$) to $E = \infty$.

Now

$$\int_0^{\infty} x^{\frac{1}{2}} \exp(- x) dx = \frac{1}{2}\sqrt{\pi}$$

and

$$n_i = \frac{1}{2}\beta_e \pi^{\frac{1}{2}} (k_B T)^{1\frac{1}{2}} \exp(- [E_g - E_f]/k_B T)$$

$$= N_c \exp(- [E_g - E_f]/k_B T \tag{4.1}$$

where $N_c [\equiv 2(2\pi m_e^* k_B T/h^2)^{1\frac{1}{2}}]$ is known as the effective density of states in the conduction band, and is about $3 \times 10^{25}/m^3$ for silicon. The difficulty with (4.1) is that we do not know where the fermi level is, but this can be ascertained by equating the expressions for n_i and p_i.

So, next to find the hole concentration, which must be the same as the electron concentration. This is given by

$$p_i = \int_{-\infty}^{0} D(E)_h [1 - F(E)] dE$$

since the probability of a hole is $1 - $ (probability of an electron), which is $1 - F(E)$. The top limit is the top of the valence band, the same as the bottom limit for n_i. This integral can be shown (see problem 4.1) to be

$$p_i = \frac{1}{2}\beta_h \pi^{\frac{1}{2}} (k_B T)^{1\frac{1}{2}} \exp(- E_f/k_B T)$$

$$= N_v \exp(- E_f/k_B T) \tag{4.2}$$

where $N_v (\equiv 2[2\pi m_h^* k_B T/h^2]^{1\frac{1}{2}})$ is the effective density of states in the valence band and m_h^* is the effective mass for holes. Since $n_i = p_i$, we must have

$$N_c \exp(- [E_g - E_f]/k_B T) = N_v \exp(- E_f/k_B T)$$

or

$$\ln N_c + E_f/k_B T - E_g/k_B T = \ln N_v - E_f/k_B T$$

so that

$$2E_f = E_g + k_B T \ln(N_v/N_c)$$

But

$$\ln(N_v/N_c) = \ln(m_h^*/m_e^*)^{1\frac{1}{2}}$$

giving

$$E_f = \tfrac{1}{2}E_g + \tfrac{3}{4}k_B T \ln(m_h^*/m_e^*) \qquad (4.3)$$

At normal temperatures $k_B T \ll E_f$ so that

$$E_f = \tfrac{1}{2}E_g$$

the fermi level lies in the middle of the band gap, and is independent of temperature. Substituting for E_f into (4.1) produces the important result

$$n_i = N_c \exp(- E_g/2k_B T)$$

or

$$n_i = N_c \exp(- E_i/k_B T) \qquad (4.4)$$

where E_i is the intrinsic fermi level, $\tfrac{1}{2}E_g$. The electron behaves as a classical particle which starts off from the middle of the forbidden gap.

Multiplying (4.1) and (4.2) together gives

$$n_i p_i = N_c N_v \exp(- E_g/k_B T) \qquad (4.5)$$

and at constant temperature (4.5) becomes

$$n_i p_i = n_i^2 = \text{constant}$$

which is called the *Law of Mass Action for Semiconductors*. Just like the Law of Mass Action in chemistry, it says that if you increase the concentration of one reactant, you depress the concentration of the other so that their product is constant. Equation (4.5) is independent of the position of the fermi level, and it also applies to extrinsic or doped semiconductors.

In an intrinsic semiconductor the conductivity may be found from

$$\sigma = n_i e \mu_e + p_i e \mu_h \qquad (4.6)$$

where e is 1.6×10^{-19} C, μ_e is the electron mobility and μ_h is the hole mobility. If we substitute into (4.6) for n_i (using the fact that $n_i = p_i$ also), we get

$$\sigma = e(\mu_h + \mu_e)N_c \exp(- E_g/2k_B T)$$

Although the mobilities are temperature dependent (and often go as $T^{1\frac{1}{2}}$), this dependency is offset by the temperature dependence of N_c (which goes as $T^{-1\frac{1}{2}}$), so the exponential term dominates the conductivity with the result that a plot of $\ln \sigma$ against T^{-1} has a slope of $-\frac{1}{2}E_g/k_B$. Figure 4.2 shows some experimental results for germanium and silicon from which their energy gaps can be found (see problem 4.2). Table 4.1 lists band gap energies produced by this means, as well as by optical absorption methods.

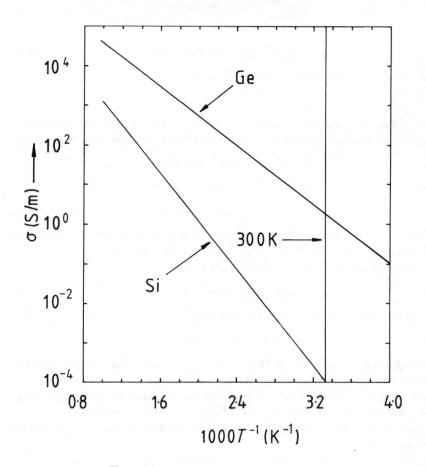

Figure 4.2 σ *against* $1/T$ *for silicon and germanium*

4.2 Extrinsic Conduction in Semiconductors

Because the concentrations of holes and electrons are the same in intrinsic semiconductors, these materials are of little practical use: what is required

Table 4.1 Some Band Gap Energies at 300 K

Material	Gap Energy (eV)	Group in Periodic Table	λ_g (nm)
C	5.4^i	IV	230
Si	1.11^i	IV	˄1110
Ge	0.66^i	IV	1880
β-SiC	2.3^i	IV–IV	540
α-SiC	2.86^i	IV–IV	434
AlP	2.45^i	III–V	497
AlAs	2.17^i	III–V	573
AlSb	1.60^i	III–V	776
GaP	2.24^i	III–V	555
GaAs	1.42	III–V	875
GaSb	0.70	III–V	1775
InP	1.35	III–V	978
InAs	0.36	III–V	3450
InSb	0.17	III–V	7310
CdS	2.42	II–VI	513
CdSe	1.74	II–VI	714
CdTe	1.44	II–VI	863

^{i} Indirect gap.
Notes: λ_g is the wavelength of a photon of energy, E_g.

If $\lambda_g > \sim 700$ nm, the material is opaque.

is a semiconductor with a *controlled* amount of one type of charge carrier. The usual means of producing a preponderance of one type of carrier is by *doping*. Take silicon as an example: it is in group IV of the Periodic Table, has four outer valence electrons, and forms covalent bonds with four other silicon atoms. In each bond two electrons, one from each atom, are shared as in figure 4.3. The bonds have been drawn in two dimensions, though

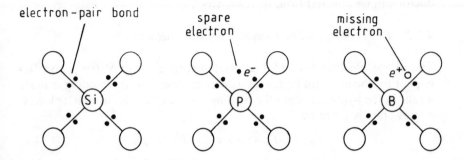

Figure 4.3 *Dopant bond-formation in silicon*

they are arranged to point to the corners of a regular tetrahedron actually. If an atom of silicon is replaced by one of the neighbouring group V elements with five outer electrons, the formation of four covalent bonds will leave a spare electron, only loosely attached to the parent atom, which is readily promoted to the conduction band as a free electron. Since no hole is left behind by this process, an excess of negative charge carriers is produced and the silicon is said to be *n-type*. In n-type silicon, electrons are said to be the *majority carriers* and holes are said to be the *minority carriers*. With usual dopant concentrations, the concentration of majority carriers is very much greater than the concentration of minority carriers, though their product is always constant by the law of mass action.

Were a group III instead of group V element to replace a silicon atom, formation of four covalent bonds would leave one of them short of an electron, producing a hole in the valence band. This hole would be only weakly bound to its parent atom, and would conduct in an electric field. There would be more holes than conduction electrons in this material, which is said to be *p-type*. In p-type material, holes are the majority carriers and electrons the minority.

4.2.1 Compensation

It is possible to have present in a semiconductor dopant atoms of both types at once. For example, suppose a p-type sample of silicon has some phosphorus diffused into it. Phosphorus tends to give up electrons to the conduction band to make the material n-type, but the product of hole and electron concentrations must remain constant by the law of mass action, so the hole concentration is reduced. If just enough phosphorus is added to make hole and electron concentrations equal, then $n = n_i$ and $p = p_i$ and the material behaves electrically as if it were intrinsic: it is said to be *compensated*. In this case the compensation is complete, though it may be partial. When both types of dopant are present, the resistivity always increases. Neither the purity nor the dopant concentration of a semiconductor can be inferred from its resistivity.

4.2.2 The Fermi Level in Extrinsic Semiconductors

The excess electron (or hole) associated with a group V (or group III) dopant in silicon is still bound to its parent atom, but the binding is fairly weak. In the hydrogen atom the binding energy of an electron in its lowest energy state is given by

$$E_I = -me^4/8\epsilon_0^2h^2 \tag{4.7}$$

where ϵ_0 is the permittivity of a vacuum. Substitution of the usual values for m etc. into (4.7) leads to $E_I = -13.6$ eV: the electron is quite tightly bound to the atom. However, assuming that the surplus electron in n-type material adopts hydrogen-like orbitals, we can replace ϵ_0 by ϵ ($= \epsilon_r\epsilon_0$), the permittivity of the medium, in this case silicon, for which the relative permittivity, ϵ_r, is 12.

The binding energy is thus reduced by a factor of ϵ_r^2 making E_I about 0.1 eV, roughly the magnitude measured for dopant ionization energies, as can be seen in table 4.2. (We should also replace m by m^*, the effective mass, in (4.7), but as the calculation is only a rough one . . .)

Table 4.2   Dopant Ionization Energies

Semiconductor	Dopant (type)	Ionization Energy
Ge	B (p)	0.010
	Al (p)	0.010
	Ga (p)	0.011
	P (n)	0.012
	As (n)	0.013
	Sb (n)	0.0096
Si	B (p)	0.045
	Al (p)	0.057
	Ga (p)	0.065
	P (n)	0.044
	As (n)	0.055
	Sb (n)	0.039

Notes: Ionization energies in eV from Arrhenius plots.
The ionization energies found optically differ a little.

The energy-band diagram for an extrinsic semiconductor looks as in figure 4.4, and the electron and hole concentrations can be calculated in much the same way as for an intrinsic material. The problem once more is to find where the fermi level is. Under normal operating conditions for a semiconductor device, the concentrations of electrons and holes are

$$n = N_D^+ \qquad \text{n-type material}$$

and

$$p = N_A^- \qquad \text{p-type material}$$

where N_D^+ is the number of ionized donor atoms/m^3 and N_A^- is the number of ionized acceptor atoms/m^3. Taking the zero for energy at the top of the valence band, the electron concentration at the bottom of the conduction band is

$$n = N_c F(E_g) \qquad (4.8)$$

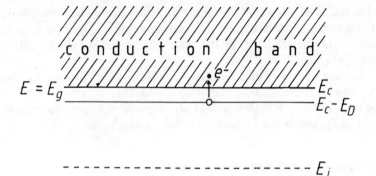

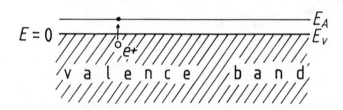

Figure 4.4   *The position of dopant energy-levels in a semiconductor*

And the hole concentration at the top of the valence band is

$$p = N_v[1 - F(E_v)] \qquad (4.9)$$

where the symbols have their previous meanings. The number of ionized atoms is given by the number of dopant atoms, N_A or N_D, times the probability of the electron or hole having the necessary energy. For ionization of the acceptor atoms, the necessary electron from the valence band has to acquire energy E_A and for ionization of the donors a hole has to have energy $(E_g - E_D)$. N_A^- and N_D^+ are therefore given by

$$N_A^- = N_A F(E_A) \qquad (4.10)$$

and

$$N_D^+ = N_D[1 - F(E_g - E_D)] \qquad (4.11)$$

As $n = N_D^+$ and $p = N_A^-$, approximately, then from (4.9) and (4.10)

$$p = N_A^- = N_v[1 - F(E_v)] = N_A F(E_A) \qquad (4.12)$$

and from (4.8) and (4.11)

$$n = N_D^+ = N_c F(E_g) = N_D[1 - F(E_g - E_D)] \qquad (4.13)$$

Equations (4.12) and (4.13) are readily, if tediously, soluble for p, n and E_f. However, some simplification is possible for the usual dopant concentration ranges, in which $N_A \ll N_v$ and $N_D \ll N_c$. Considering equation (4.12), $E_v = 0$, as the top of the valence band is taken as the zero for energy, and $1 - F(0) \approx \exp(- E_f/k_B T)$. Then if $E_A \ll E_f$, $F(E_A) \approx 1$, so that (4.12) becomes very simply

$$N_v \exp(- E_f/k_B T) \approx N_A$$

or

$$E_f \approx k_B T \ln(N_v/N_A) \tag{4.14}$$

For silicon N_v is about $3 \times 10^{25}/m^3$ at 300 K and if N_A is $10^{22}/m^3$ (an acceptor concentration of about 0.2 ppm) we find

$$E_f = 0.025 \ln(3000) = 0.2 \text{ eV}$$

taking $k_B T = 0.025$ eV. An exact solution of (4.12) with $E_A = 0.05$ eV gives $E_f = 0.17$ eV. Increasing the concentration of donor atoms moves the fermi level closer to the valence band and decreasing the concentration moves the fermi level towards the centre of the band gap, the intrinsic fermi level.

Equation (4.13) may be solved in a like manner giving

$$E_f \approx E_g - k_B T \ln(N_c/N_D) \tag{4.15}$$

In this case, too, the fermi level lies near the dopant energy level, close to the band edge. These results may be summarized:

- *In an intrinsic semiconductor the fermi level lies at the centre of the band gap.*
- *In a doped semiconductor the fermi level lies near the dopant level.*

Increasing the temperature in a doped semiconductor moves the fermi level towards the centre of the band gap, in other words the material tends to become intrinsic as might be expected. The overall picture of the fermi level's position as dopant concentration and temperature change is shown in figure 4.5. If $\ln (n)$ or $\ln (p)$ is plotted against T^{-1} for a sample of silicon the result is as shown in figure 4.6, from which we can identify three regimes:

(1) At low temperatures, below $\sim$ 100 K, the carrier concentration goes as $\exp(- E_I/k_B T)$, where E_I is the dopant's ionization energy. The dopant atoms are not fully ionized in this region as $k_B T \ll E_I$. This is the impurity or extrinsic regime. The fermi level is at the dopant level, that is, at E_A or $(E_g - E_D)$.

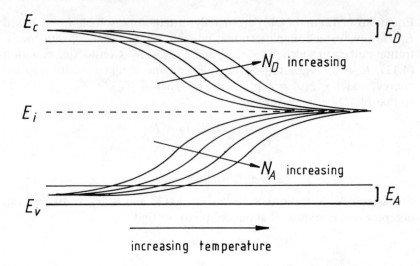

Figure 4.5 *The change of E_f with N_A, N_D and T in a semiconductor*

(2) Above 100 K and up to ~ 450 K the carrier concentration does not vary much with temperature. In this region the dopant atoms are fully ionized and the concentration of thermal holes and electrons from the silicon (the 'lattice') is much less than that of the dopant atoms. This is known as the saturation or exhaustion regime. The fermi level lies near the dopant level, but towards the centre of the band-gap.

(3) Above ~ 450 K the carrier concentration varies as $\exp(-E_g/2k_BT)$. The thermally generated carriers from the lattice are in larger concentrations than those from the dopant. This is the intrinsic regime. The fermi level lies in the middle of the band gap.

Devices are usually operated in the second, saturation regime. Because *n* or *p* is constant in the exhaustion region, any conductivity variation is mainly due to changes in mobility with temperature. Near room temperature the mobility in silicon declines with increasing temperature in the usual doping range of device material; this is why the TCR is positive and not negative as one might expect.

4.3 p–n Junctions

Semiconductor devices make a great deal of use of p–n junctions, regions where p- and n-type semiconductor material are joined. There is a variety

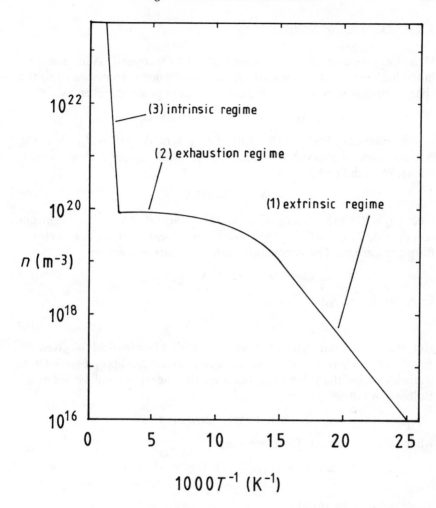

Figure 4.6 *Graph of ln(n) against 1/T in silicon*

of techniques for producing them (as we shall see), but first we shall look at the properties of ideally abrupt junctions. The current in p-type material is mainly carried by the majority carriers – holes; and in n-type material, by electrons. If n- and p-type material are brought into contact, current flows across the junction even though no external potential is applied. Minority carriers are injected into each type of material, and these recombine with majority carriers after a time, called the *minority carrier lifetime*, or just the *lifetime*. The Einstein relation is useful for calculating how far these minority carriers travel before recombining.

4.3.1   The Einstein Relation

If the local concentration of minority carriers is increased, as for example, in the base region of a transistor, there is a diffusion current caused by the charge carriers moving down the concentration gradient, given by

$$J_D = -eD\nabla N = -eD dN/dx$$

in one dimension, where the diffusion current density is J_D, N is the concentration of minority carriers, and the diffusivity (or diffusion coefficient), D, is defined by

$$N v_d = -D\nabla N$$

where v_d is the drift velocity of the minority carriers. The diffusion current sets up a field, $\mathscr{E}$, such that the conduction current it produces cancels the diffusion current. The conduction current density is given by

$$J_c = \sigma\mathscr{E} = Ne\mu\mathscr{E}$$

Equating J_c and J_D

$$Ne\mu\mathscr{E} = -eD dN/dx \qquad (4.16)$$

But the energy an electron acquires in an electric field is given by $E = -e\mathscr{E}x$, where x is the distance gone, so we can apply Fermi–Dirac statistics to find the value of N, but since the energy is small we are in the Boltzmann tail and

$$N = C\exp(-e\mathscr{E}x/k_B T)$$

where C is a constant. Differentiating:

$$dN/dx = (-e\mathscr{E}/k_B T)C\exp(-e\mathscr{E}x/k_B T)$$
$$= -e\mathscr{E}N/k_B T$$

Substituting in (4.16) gives

$$Ne\mu\mathscr{E} = e^2 D\mathscr{E}N/k_B T$$

or

$$\mu = De/k_B T$$

which is the *Einstein relation*. It is useful for finding D once the mobility is known (see section 4.6).

From the diffusivity, D, it is possible to estimate how far the minority carriers will diffuse before recombining with the majority carriers, which is the *diffusion length*, l_d. It is found from

$$l_d = \sqrt{(D\tau)}$$

where τ is the *lifetime* of the minority carriers (not to be confused with the relaxation time, also called τ). It is possible also to measure τ (again, see section 4.6), so that l_d can be calculated. Typically, $\tau \approx 1$ μs in silicon devices, and with $\mu \approx 0.1$ m^2/V/s, the Einstein relation gives $D \approx 2.5 \times 10^{-3}$ m^2/s, so that $l_d \approx 50$ μm. This is big compared with distances in transistors, which are of the order of a few μm at most.

4.3.2   The Depletion Region

When a p–n junction is established, the electrons and holes either side of the junction must come into equilibrium. For conduction electrons to be in equilibrium in both p- and n-type material, the fermi level must be the same in both. Now we have seen that in p-type material the fermi level is near the valence band maximum, that is $E_{fp} \approx 0$, while in n-type material it is near the conduction band minimum, that is $E_{fn} \approx E_g$. So the energy levels on the n-type side of the junction are depressed at equilibrium, or the p-type side's are raised, by an amount, eV_{bi} equal to the original difference in fermi levels, as shown in figure 4.7(a). E_{fn} and E_{fp} are approximately given by equations (4.14) and (4.15), from which

$$eV_{bi} = E_{fn} - E_{fp}$$
$$= E_g - k_B T \ln(N_c/N_D) - k_B T \ln(N_v/N_A) \qquad (4.17)$$

By expressing E_g in terms of the intrinsic carrier concentration given in (4.5), equation (4.17) can be rewritten in the form

$$eV_{bi} = k_B T\{\ln(N_c N_v/n_i^2) - \ln(N_c/N_D) - \ln(N_V/N_A)\}$$

or

$$eV_{bi} = k_B T \ln(N_D N_A/n_i^2) = k_B T \ln(n_n p_p/n_i^2) \qquad (4.18)$$

where n_n is the concentration of electrons on the n-type side far from the junction and p_p is the concentration of holes on the p-type side far from the junction . In the case of a typical junction diode, we might find $N_A (= p_p)$ to be 10^{21}/m^3 and $N_D (= n_n)$ to be 10^{23}/m^3, and as $n_i = 1.5 \times 10^{16}$/m^3 (4.18) leads to $V_{bi} = 0.7$ V at 300 K.

The *built-in potential*, as this voltage is called, results in a field across the junction which prevents the diffusion of charge carriers beyond a region known as the *depletion region*. In this volume of material close to the junction, holes have diffused across into the n-type material from the p-type and there have combined with free electrons to leave a region with almost no free charge carriers. However, the neutralization of the electrons in the n-type material leaves a positive charge on the fixed dopant ions. The situation on the p-type side is similar, with holes being neutralized by electrons diffusing from the n-type side, leaving negatively charged, fixed dopant ions. The picture looks as in figure 4.7. Because the

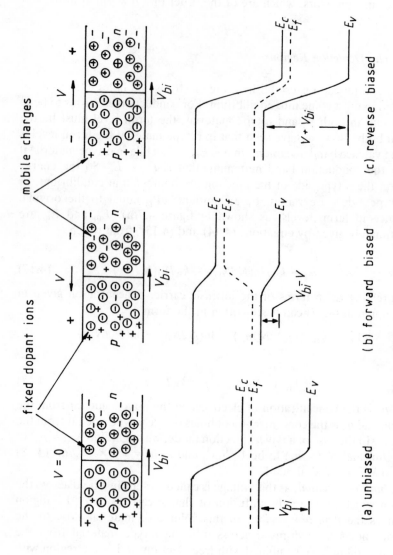

Figure 4.7   *The energy levels at a p–n junction under different bias conditions*

depletion region contains these fixed lattice charges it is sometimes called the *space charge* layer.

One can show (see problem 4.4) that V_{bi} is directly related to the band gap by

$$eV_{bi} \approx k_B T \ln(N_A N_D / N_c N_v) + E_g$$

The first term on the left is about -0.4 eV for normal doping levels in silicon at 300 K.

The width of the depletion region is readily calculated from Poisson's equation, assuming a constant space-charge density in the depletion layer and an abrupt junction – no concentration gradient of dopant in n- or p-regions – as in figure 4.8. Poisson's equation is

$$\epsilon \nabla^2 \psi = - q_v$$

ϵ is the permittivity $(= \epsilon_r \epsilon_0)$, ψ is the electrostatic potential and q_v is the charge/m^3 (we should emphasize that it is the *fixed* charge on the ionized dopant atoms that concerns us here). On the p-type side, q_v is just $-eN_A$, and on the n-type side it is eN_D. Integrating Poisson's equation in one dimension, we have on the p-type side

$$\epsilon d\psi/dx = eN_A \int dx = eN_A x + C$$

C is a constant of integration. Now at $x = -x_p$, the field, $\mathscr{E}$ $(= - d\psi/dx)$ is zero, so $C = eN_A x_p$ and

$$\epsilon d\psi/dx = eN_A(x + x_p) \tag{4.19}$$

On the n-type side,

$$\epsilon d\psi/dx = - eN_D \int dx = - eN_D x + C'$$

C' is another constant of integration, which is $eN_D x_n$, as $\mathscr{E} = 0$ at $x = x_n$, so that

$$\epsilon d\psi/dx = - eN_D(x - x_n) \tag{4.20}$$

Integrating:

$$\epsilon \int d\psi = \epsilon V_{bi} = eN_A \int_{-x_p}^{0} (x + x_p)dx + (- eN_D) \int_{0}^{x_n} (x - x_n)dx$$

$$V_{bi} = \frac{e}{2\epsilon}(N_A x_p^2 + N_D x_n^2) \tag{4.21}$$

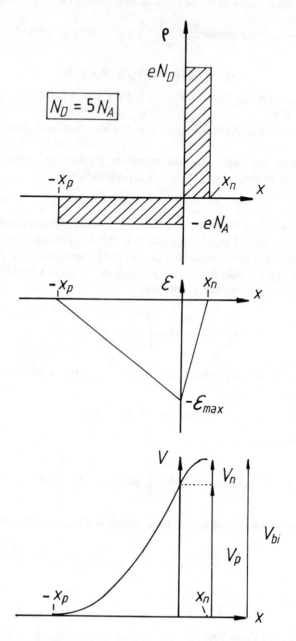

Figure 4.8   *Dopant concentration,$\mathcal{E}$ and V in the depletion region*

Now we can find the width of the depletion region, which is

$$w = x_p + x_n \qquad (4.22)$$

Charge conservation requires the charges to be equal each side of the junction, that is

$$N_A A x_p = N_D A x_n \qquad (4.23)$$

taking A to be the cross-sectional area of the junction. Substitution for x_p into (4.21) gives x_n, and substitution for x_n in (4.21) gives x_p, which can in turn be put into (4.22) to yield

$$w^2 = (2\epsilon V_{bi}/e)([N_A + N_D]/N_A N_D)$$

For silicon $\epsilon_r = 12$, so $\epsilon = 12\epsilon_0$ and if $N_A = 10^{21}/m^3$ and $N_D = 10^{23}/m^3$, making $V_{bi} = 0.7$ V as before, then $w \approx 1$ μm, a typical value. When the dopant concentration ratio is high, the depletion region lies almost entirely on the more lightly doped side as seen from (4.23). In this case we can write

$$w \approx \sqrt{(2\epsilon V_{bi}/Ne)} \qquad (4.24)$$

where N is the dopant density in the more lightly doped side. The error in (4.24) is only 0.5 per cent in this example.

The field in the depletion region may be calculated from equations (4.19) and (4.20) as $\mathscr{E} = - d\psi/dx$. It attains its maximum value at the junction, where $x = 0$. It is easy to show (see problem 4.3) that when the doping ratio is high

$$\mathscr{E}_{max} \approx 2V_{bi}/w$$

$\mathscr{E}_{max}$ can reach very high values, 1 MV/m or more, since V_{bi} is about 1 V and w about 1μm.

The depletion layer width is voltage-dependent, and the formulae above apply only when the applied voltage is zero. When the applied voltage is not zero the base width is

$$w \approx \sqrt{(2\epsilon[V_{bi} - V]/Ne)} \qquad (4.25)$$

where V, the applied voltage, is positive for forward-biased junctions and negative for reverse-biased ones.

The junction capacitance, an important parameter in device operation because it sets limits to the frequency of operation, is readily calculated from the depletion layer width. Recall the capacitance of a parallel plate capacitor is given by

$$C = \epsilon A/w$$

so that

$$C/A = C_A = \sqrt{(\tfrac{1}{2}\epsilon Ne/[V_{bi} - V])} \qquad (4.26)$$

where ϵ is the permittivity of the medium, A is the plate area and w is the distance between the plates, given by (4.24). For the dopant levels in the above case, C_A, the capacitance per unit area, works out to about 100 pF/mm^2, at zero bias. Varactor diodes having voltage-variable capacitance use this principle.

4.3.3 The Rectifier Equation

The p–n junction can rectify current: in forward bias (that is, when the p-side is more positive than the n-side) it conducts, while in reverse bias it allows little current to flow.

At equilibrium, without any applied potential, there are equal and opposite flows of current across the junction. Electrons from the n-region diffuse into the p-region and recombine with holes, hence the name electron recombination current, I_{er}. There are also thermally generated electrons (not very many) in the p-region, which drift across the junction into the n-region, constituting the electron generation current, I_{eg}. Charge conservation requires the two electron currents to be equal at equilbrium:

$$I_{er} = I_{eg}$$

The same considerations apply to holes

$$I_{hr} = I_{hg}$$

so there are four current terms.

When the junction is reverse biased, the voltage across the junction is increased by the bias potential, V, so the electrons have to acquire further energy, eV, to surmount the increased barrier. Thus I_{er} is reduced by the Boltzmann term, $\exp(-eV/k_BT)$:

$$I_{er}(-V) = I_{er}\exp(-eV/k_BT)$$

However, the electron generation current is not affected, because the electrons are always at the top of the potential barrier and raising it does not affect their thermal generation, so

$$I_{eg}(-V) = I_{eg}$$

The generation current is now much larger than the recombination current, but is still very small as the density of thermally generated carriers is very small.

When the junction is forward biased, the electrons have a potential barrier (see figure 4.7) to surmount which is reduced by the bias voltage, V, so that the recombination current is increased by $\exp(eV/k_BT)$:

$$I_{er}(+V) = I_{er}\exp(eV/k_BT)$$

The generation current is unchanged for the same reason as before. Holes are affected by the bias in the same way as the electrons, but flow in the opposite direction, so that in forward bias

$$I_{hr}(+\ V) = I_{hr}\exp(eV/k_BT)$$

The four currents add algebraically:

$$I(V) = I_{er}(V) - I_{eg}(V) + I_{hr}(V) - I_{hg}(V)$$
$$= (I_{er} + I_{hr})\exp(eV/k_BT) - I_{eg} - I_{hg}$$

as $I_{er} = I_{eg}$ etc., and so

$$I(V) = I_s[\exp(eV/k_BT) - 1] \tag{4.27}$$

(4.27) is the rectifier, or ideal diode, equation. I_s ($= I_{er} + I_{hr}$), the reverse saturation current, is $\approx 1\ \mu A/mm^2$ in germanium diodes and $\approx 100\ pA/mm^2$ in silicon diodes at 300 K. Figure 4.9 shows graphs of equation (4.27) for germanium and silicon at room temperature.

The current flowing through the depletion region is assumed in this derivation to be entirely due to carriers from outside the depletion

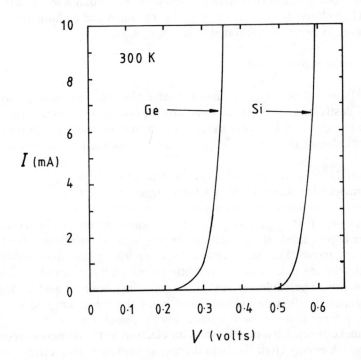

Figure 4.9 *I(V) graphs for p–n junctions in Ge and Si*

region – thermally generated carriers within the depletion region play no part in conduction. This is almost true for germanium, but not for silicon or other materials with low intrinsic carrier concentrations. In these cases the ideal diode equation needs some modification, but it is still of exponential form.

At room temperature, (4.27) reduces approximately to

$$I = I_s\exp(40V)$$

so that

$$dI/dV = 1/R_{dyn} = 40I_s\exp(40V)$$

or

$$R_{dyn} = (40I_s)^{-1}\exp(-40V) \tag{4.28}$$

R_{dyn} is the *dynamic resistance* of the diode. For a small silicon signal diode, I_s may be 1 pA and R_{dyn} is then about 1 Ω for a forward bias of 0.6 V and a forward current of 25 mA. If the forward bias drops to 0.4 V, then R_{dyn} is about 3 kΩ and the current falls to 10 μA. For this reason silicon diodes are said to require a forward bias of about 0.6 V to conduct, but they are, of course, conducting even with much smaller biases; but not much current flows. The voltage drops associated with germanium and gallium arsenide diodes may likewise be calculated in problem 4.5.

4.3.4 Junction Breakdown

p–n junctions will not support infinite reverse bias, but will conduct large, possibly destructive, currents at some critical *breakdown voltage*. Breakdown is made use of in Zener diodes, which may be used to control the voltage at a point in a circuit. There are two breakdown mechanisms:

(1) Tunnelling – at reverse voltages less than about $5E_g/e$.
(2) Avalanche breakdown – at reverse voltages above $5E_g/e$.

Tunnelling – first suggested by Zener – is a quantum effect which occurs when the doping levels at the junction are very high, so high in fact that the fermi level moves into the valence band on the p-side and into the conduction band on the n-side, a condition in which the material is said to be *degenerate*. The depletion region is very thin (about 10 nm) in degenerate material, allowing electrons (or holes the other way) to tunnel between valence and conduction bands under a small reverse bias.

Avalanche breakdown occurs when an electron in the depletion region gains enough energy from the field therein to liberate hole–electron pairs on collision with the lattice atoms. The electron gains energy $e\mathscr{E}l_m$ from the field, where l_m is the mean free path, and to produce a hole–electron pair this must be larger than the band gap, at least (in fact, about $4E_g$). The critical field in the depletion region, $\mathscr{E}$, is of the order of V_b/w, where V_b is the breakdown voltage and w is the depletion layer width, so that

$$e\mathscr{E}l_m = eV_bl_m/w \approx 4E_g$$

$$V_b \approx 4wE_g/el_m$$

For $w \approx 1$ μm, $E_g = 1.1$ eV (silicon) and $l_m \approx 20$ nm, then $V_b \approx 220$ V. To achieve high breakdown voltages requires large depletion-layer widths and hence low doping levels and vice versa.

4.4 The Bipolar Junction Transistor

A bipolar junction transistor has two p–n junctions arranged back-to-back with a small region separating them called the *base*. The term *bipolar* is used because two kinds of charge carrier are involved in its operation.

Consider a p–n–p transistor, as in figure 4.10, connected in common base configuration; that is, the base is common to both the input (*emitter*) and output (*collector*) circuits. With the applied voltages shown, the emitter–base p–n junction is forward biased, so holes (the majority carriers in the emitter) can cross from emitter to base readily. In the n-type base region, the holes injected from the emitter are minority carriers and should recombine with electrons to form the base current. However, the base width is much less than $\sqrt{(D\tau)}$, the distance the minority carriers will go before recombination on average, so the holes reach the reverse-biased base–collector junction. But holes in the base are minority carriers, so they will easily cross the reverse-biased n–p junction between base and collector (the reverse bias prevents holes crossing easily from collector to base), and so will appear in the collector circuit as the collector current. There is also an electron current in the opposite direction to the hole current (but constituting a conventional current in the same sense as the hole current). Figure 4.10 shows these currents.

The common-base current gain (α) is just less than unity, as some holes have recombined in the base to give a small base current, but the voltage gain is large. As an approximation

$$\alpha = 1 - \tfrac{1}{2}(W/l_d)^2 \tag{4.29}$$

where W is the base-width and l_d is the diffusion length of the minority carriers (in this case holes) in the base. Suppose the hole mobility is 0.05 m²/V/s, then the diffusivity, D, from the Einstein relation is 1.3×10^{-3} m²/s. If the lifetime of minority carriers in the base is 0.3 μs, l_d becomes 20 μm and α is 0.999 when the base-width is 1 μm. These are typical values for a silicon device, though α is usually rather less than this

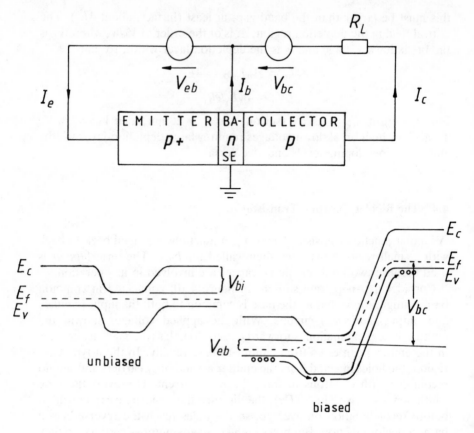

Figure 4.10   *Energy levels in a p–n–p transistor*

value: about 0.98–0.998. Equation (4.29) shows the importance for high transistor gain of small base-widths and large diffusion lengths, which implies high lifetimes and high carrier mobilities. The common emitter current gain, β, is given by

$$\beta = \alpha/(1 - \alpha)$$

so that α should be as close to unity as possible. β is usually in the range 50–500.

4.5   The MOSFET

Nowadays most integrated circuits are assemblies of MOS (Metal–Oxide–Silicon) devices such as the MOSFET (MOS Field Effect Transistor), which are somewhat easier to fabricate than bipolar devices. Because its operation involves only one type of charge carrier, it is known as a unipolar device: it is simpler to understand as well as to make.

Figure 4.11 shows an n-channel MOSFET in section. The substrate material is p-type silicon (boron-doped), into which shallow diffusions of phosphorus or arsenic are made to produce heavily doped (n+) regions known as the source and drain. Metallic contacts are made directly to the n+ regions, while the region between source and drain, known as the gate, is not contacted directly, but is insulated from the substrate by a layer of silicon dioxide produced by oxidation of the substrate (see chapter 5). This is the origin of another acronym – IGFET (Insulated Gate FET). The gate is used to control the flow of current between source and drain by means of an induced channel beneath the gate electrode.

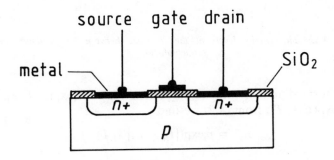

Figure 4.11 *An n-channel MOS transistor*

To understand the formation of the channel under the gate electrode we must look at the carrier concentrations at the oxide–semiconductor interface. The energy-band picture is as in figure 4.12. The electrostatic potential in the bulk of the semiconductor is ψ_0, ψ_s is the surface potential, or the voltage applied to the gate, more or less, ψ_Δ is the difference between E_i (the intrinsic fermi level) and E_f (the fermi level in the semiconductor, in this case near E_v, as the substrate is p-type). The surface charge densities are

$$\left.\begin{array}{l} n_s = n_i \exp[e(\psi_s - \psi_\Delta)/k_B T] \\ p_s = p_i \exp[e(\psi_\Delta - \psi_s)/k_B T] \end{array}\right\} \tag{4.30}$$

When $\psi_s = \psi_\Delta$, the carrier concentrations are intrinsic and $E_i = E_f$ at the surface. If $\psi_s > \psi_\Delta$, then $n_s > p_s$ and the surface layer is n-type; an *inversion layer* is said to have formed. ψ_Δ is determined by the band gap of the material and the doping level, N_A. Normally the channel is not

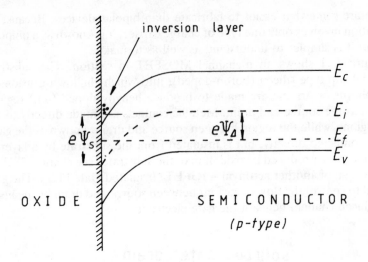

Figure 4.12 *The energy levels at the interface between an insulator and a semiconductor*

properly formed until $n_s = n_s^+ \approx N_A$, and as $N_A \approx N_v\exp(-E_f/k_BT)$ and $N_v = p_i\exp(E_i/k_BT)$, from (4.2) we find

$$n_s^+ = p_i\exp([E_i - E_f]/k_BT)$$

$$= p_i\exp(e\psi_\Delta/k_BT)$$

Comparison with (4.30) shows that, for strong inversion

$$\psi_s^+ = 2\psi_\Delta = 2(k_BT/e)\ln(N_A/p_i)$$

ψ_s^+ is about 0.7 V in silicon.

The potential applied to the gate is split between the gate insulation layer and the semiconductor, so that

$$V_g = V_i + \psi_s$$

and

$$V_i = q/C_i = q_s/C_{is}$$

C_i is the gate capacitance, C_{is} is the gate capacitance/m^2 and q_s is the surface charge/m^2 in the depleted semiconductor:

$$q_s = eN_Aw$$

where w is the width of the surface inversion layer, given by (4.24), with $V_{bi} = \psi_s$, so that

$$V_i = \sqrt{(2eN_A\epsilon_s\psi_s)}/C_{is}$$

and

$$C_{is} = \epsilon_i/d$$

where ϵ_s and ϵ_i are the permittivities of the semiconductor and insulator respectively, d is the insulator thickness. Thus the *threshold voltage* of the MOSFET will be

$$V_T = (d/\epsilon_i)\sqrt{(2N_A e\epsilon_s \psi_s^+)} + \psi_s^+ \qquad (4.31)$$

Typically, $N_A \approx 10^{22}/\text{m}^3$, $d \approx 50$ nm, $\epsilon_r \approx 3$ for the insulation (that is, $\epsilon_i = 3\epsilon_0$) and $\epsilon_r \approx 12$ for silicon (that is, $\epsilon_s = 12\epsilon_0$), so if ψ_s^+ is 0.7 V, V_T is about 1.6 V. Equation (4.31) shows that it is important to keep the gate insulation thin, and thicknesses as low as 10 nm are not uncommon.

4.6 Measurement of Semiconductor Properties

Knowledge of the properties of semiconductor material is very important for device operation. Measurement of the properties of a material is usually termed *characterization*. The most important properties are conductivity, conductivity type, Hall coefficient, carrier concentration, mobility, lifetime, effective mass and band gap. Of course, in most cases the latter is known, but new materials are being developed that rely on precise control of the band gap, so in some cases its measurement is a routine necessity.

4.6.1 Conductivity and Type

The conductivity is readily measured on a relatively large, regularly shaped piece of material, but most semiconductor material is processed in the form of thin slices. The easiest way to measure the conductivity of these is by use of a four-point probe (ASTM test method F84), as in figure 4.13. The probe is pressed into contact with the surface of the material enclosed in a dark box to prevent photo-electric carrier generation, and a controlled current is passed between the outer pair of probes while the voltage between the inner pair is recorded. If the probe spacing and thickness of the specimen are known, then the ratio of I/V can be converted into a conductivity. The accuracy is about 2 per cent.

Conductivity type can be determined on a slice of material by using a hot probe (ASTM method F42) as in figure 4.14. The hot probe can be as crude as a soldering iron, but more accuracy is obtained from specially-built ones. The majority carriers at the hot probe diffuse out to the cold probe, so setting up a voltage. When the carriers are electrons their outward diffusion leaves the hot probe positive with respect to the cold,

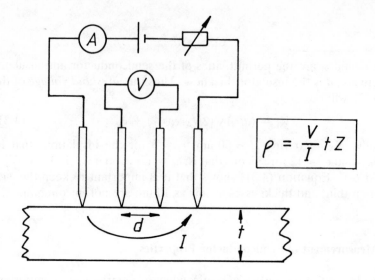

Figure 4.13   *The four-point probe*

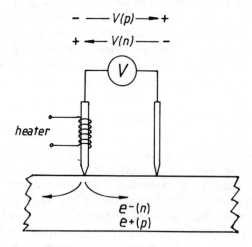

Figure 4.14   *The hot probe*

and vice versa for p-type material. The method works for resistivities up to 10 Ωm.

4.6.2   *The Mobility and Lifetime*

Although it is possible to calculate the mobility of majority carriers from the Hall coefficient and conductivity, the mobility of minority carriers can be measured by means of the Haynes–Shockley experiment, which is shown in figure 4.15. Minority carriers are injected by flashing a high-intensity light onto the surface of the semiconductor, and these drift down the potential gradient to a probe connected to an oscilloscope. The time taken for the peak amplitude to appear at the oscilloscope, t, is x/v_d which is $x/\mu\mathscr{E}$, where $\mathscr{E} = V/l$, so $\mu = xl/Vt$. The lifetime can be estimated from the rate of decay of the peak amplitude with time, since

$$V_p = V_0\exp(- t/\tau)$$

where V_0 is the peak amplitude at $t = 0$ and V_p is the peak amplitude at time t.

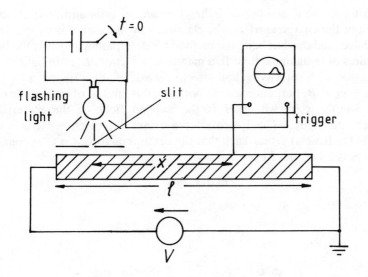

Figure 4.15   *The Haynes–Shockley experiment*

4.6.3   The Hall Coefficient

This has been discussed in chapter 2. In semiconductors containing significant concentrations of both types of carriers, that is in a compensated material, if $\mu_n/\mu_p(\equiv b)$ is the ratio of electron to hole mobility, the Hall coefficient is

$$R_H = \frac{1}{e} \frac{p - nb^2}{(p + nb)^2} \tag{4.32}$$

Suppose $b = 5$, $p = 10^{23}/m^3$ and $n = 10^{22}/m^3$, then R_H is $- 4 \times 10^{-5}$ m^3/C, so that the material appears to be n-type even though there are ten times as many holes per unit volume as electrons.

4.6.4   The Carrier Concentration

Usually this is estimated from $\sigma = ne\mu_e$ or $\sigma = pe\mu_h$, assuming only one type of carrier is important. When compensation has occurred, then if the mobilities are known, the carrier concentrations can be found from σ and R_H (see problem 4.6).

4.6.5   The Effective Mass

While there are a number of indirect means of estimating the effective masses of the charge carriers, the classical way of doing it is by cyclotron resonance, and this has been the method for determining much of the band structures of semiconductors. In a magnetic induction, B, as in figure 4.16, an electron with velocity, v, and effective mass, m^*, experiences a Lorentz force, $- ev \times B$, perpendicular to both v and B, and so of magnitude evB. (B is simply $\mu_0 H$, where H is the applied field, if the material is non-ferromagnetic.) The Lorentz force causes the electrons to describe circular (or helical) orbits, such that the centripetal force $(m^* v^2/r)$ counterbalances the Lorentz force:

$$evB = m^* v^2/r$$

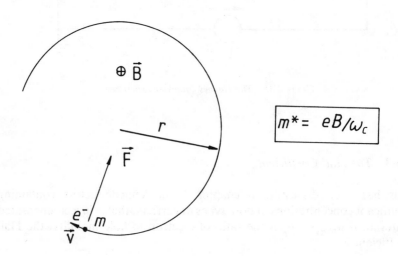

Figure 4.16   *Cyclotron resonance*

The period is given by

$$T = 2\pi r/v = 1/f_c$$

$$m^* = eB/2\pi f_c$$

where f_c is the cyclotron resonance frequency. Suppose $B = 0.5$ T and m^* is $0.5m_0$, then f_c is 28 GHz, in the microwave region. Because the electrons must complete at least 20 per cent of a circle between collisions for resonance to be observed, the experiment is usually done at low temperatures where the relaxation times are long. In a typical experiment, B is varied from 0 to 1 T at a fixed frequency of about 20–30 GHz, and the absorption peak gives the effective mass, or masses, as there are often several because of the complicated band structures of most materials (see problem 4.7).

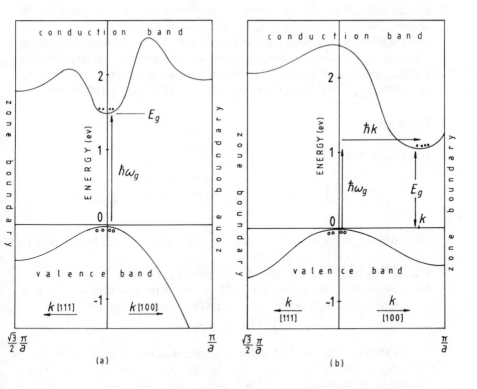

Figure 4.17 *(a) The band structure and photon production in a direct-gap semiconductor; (b) the band structure and photon production in an indirect-gap semiconductor*

4.6.6   The Energy Gap

An Arrhenius plot of conductivity against $1/T$ has a slope of $-E_g/2k_B$ in the intrinsic region, as we have seen. It is also possible to plot the absorption against wavelength in the infra-red and visible regions of the electromagnetic spectrum and look for the point where the absorption begins to rise more sharply – the *absorption edge*. In *direct gap* materials, such as gallium arsenide, this is quite sharp and the energy gap is measurable precisely, but in materials with *indirect* gaps, such as silicon and germanium, the absorption process is more complicated because of the momentum change of the electron. The picture is as in figure 4.17. In the direct gap material a photon of energy E_g promotes an electron across the gap at $k = 0$:

$$E_g = hf/e = hc/e\lambda_g$$

E_g is in eV. Here, λ_g is 830 nm (that is, near infra-red) and E_g is 1.5 eV. For an indirect-gap material, the electron from the valence band has to acquire additional momentum, $\hbar k$, to make the transition from valence band maximum to conduction band minimum. This additional momentum comes from a lattice vibrational phonon, which must have exactly the right k-value, thus lowering the probability of the photon's exciting the electron across the gap. The absorption edge is then much less sharply defined.

Problems

4.1.   Show that

$$\int_{-\infty}^{0} D(E)_h[1 - F(E)]dE = N_v\exp(-E_f/k_BT)$$

where $D(E)_h \equiv (2\pi^2)^{-1}(2m_h/\hbar^2)^{\frac{1}{2}}V(-E)$, $N_v \equiv 2(2\pi m_h k_B T/h^2)^{\frac{1}{2}}$ and $F(E)$ is the fermi function.

4.2.   From figure 4.2 find the energy gaps for silicon and germanium.

4.3.   Show that the maximum field in the depletion region of a p–n junction of high doping ratio is given approximately by

$$\mathscr{E}_{max} \approx 2V_{bi}/w$$

Estimate $\mathscr{E}_{max}$ at 300 K for a semiconductor with $N_A = 10^{22}/m^3$, $N_D = 10^{23}/m^3$, $N_c = N_v = 3 \times 10^{25}/m^3$, $\epsilon_r = 8$ and $E_g = 2.5$ eV. [Use the approximation for V_{bi} given in problem 4.4.]
[Ans. 9.4 MV/m]

4.4.   Show that

$$eV_{bi} \approx k_B T\ln(N_A N_D/N_v N_c) + E_g$$

If $N_A \approx N_v/1000$ and $N_D \approx N_c/1000$, what will be the built-in potentials for InSb, GaAs, CdS, β-SiC and diamond at 300 K? Comment.
[*Ans:* − *0.19 V, 1.06 V, 2.17 V, 1.94 V and 5.04 V*]

4.5. Use the rectifier equation (4.27) to find the forward current at 300 K for p–n junction diodes when $R_{dyn} = 1\ \Omega$. What are the associated forward voltages for diodes made from GaAs, Ge and β-SiC?
[Take I_s to be 10^{-17}, 10^{-6} and 10^{-32} A for GaAs, Ge and β-SiC respectively.]
[*Ans. 26 mA; 0.92 V, 0.26 V and 1.8 V*]

4.6. A certain semiconductor material is found to have a resistivity of 167 Ωm and a Hall coefficient of 6.1 m³/C at 300 K. If $\mu_e = 0.16$ m²/V/s and $\mu_h = 0.04$ m²/V/s, find the concentrations of electrons and holes. What would be the intrinsic conductivity at 300 K?
[*Ans. $4 \times 10^{15}/m^3$, $9.2 \times 10^{17}/m^3$; 1.94×10^{-3} S/m*]

4.7. Figure P4.7 shows the results from a hypothetical cyclotron resonance experiment at 25 GHz. Find the effective masses of all the particles in terms of the electronic rest mass, m_0.

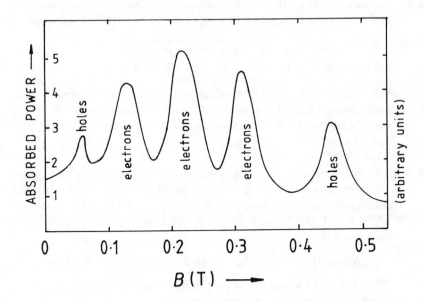

Figure P4.7 *The results of a cyclotron resonance experiment*

The line width, ΔB, at the − 3 dB points can be used to estimate the carrier collision time, τ, since the microwave absorption is proportional to the imaginary part of the wave-vector, k, where

$$k = \omega \sqrt{(\mu\epsilon)} \left[1 + \frac{j\omega_p^2}{2\omega} \frac{1}{1 + (\omega - \omega_c)^2 \tau^2} \right]$$

ω is the angular frequency, ω_c is the cyclotron resonance frequency, $j = \sqrt{(-1)}$ and μ, ϵ and ω_p are constants. Estimate the collision time for the heaviest particle.

[Ans. $0.066m_0$, $0.15m_0$, $0.24m_0$, $0.35m_0$ and $0.5m_0$; 5×10^{-11} s]

4.8. Assuming that germanium, silicon and β-SiC are identical electronically except for band gap, what will be the minimum dopant concentration required at room temperature for a p–n junction in each material to operate?

[Assume the dopant concentration must be at least a hundred times the intrinsic carrier concentration. Use table 4.1 for the band gaps. Take $N_v = N_c = 3 \times 10^{25}/m^3$.]

If the dopant concentration in a p–n junction device is $10^{23}/m^3$, what will be its maximum operating temperature when it is made from germanium, from silicon and from β-SiC?

[Ans. $8.7 \times 10^{21}/m^3$, $1.2 \times 10^{18}/m^3$, $1.5 \times 10^8/m^3$; 98°C, 356°C and 1020°C]

4.9. Measurements on the conductivity of a specimen of semiconducting material are made at several temperatures as in the table below.

$T(°C)$	300	400	500	600	700	800	900
$\sigma(kS/m)$	8.1	9.1	14.0	21	67	196	470

Estimate the temperature at which intrinsic behaviour begins. What is the energy gap of this material? At what wavelength will optical absorption begin? What colour would a single crystal be in transmitted light?

[Ans. 600°C; 678 nm]

4.10. The Hall coefficient is given by equation (4.31) when both holes and electrons are present. If $b = 10$ in (4.31) show that the Hall coefficient has its maximum value when $p = 18.1n_i$ and its minimum when $p = 1.75n_i$, where n_i is the intrinsic carrier concentration. Show also that

$$R_H(min) = -0.994/n_i e \quad \text{and} \quad R_H(max) = 0.036/n_i e$$

What values of n and p will give minimal conductivity?

[Ans. $p = 3.16n_i$; $n = n_i/3.16$]

4.11. Show that the TCR may be found from a plot of $\ln/\sigma$ against T^{-1}, and is given by

$$\text{TCR} = s/T^2 = -E_g/2k_B T^2$$

where s is the slope of the graph. Hence find the TCR for silicon at 300 K in the intrinsic regime from figure 4.6. Estimate the TCRs of intrinsic InAs and Ge at 300 K and compare with the experimental values given in table 2.1.

[The TCR may be taken as $\rho^{-1}d\rho/dT$.]

[Ans. $-$ 0.074/K; $-$ 0.023/K and $-$ 0.042/K]

5

VLSI Technology

VLSI (Very Large Scale Integration) is responsible for the very complex, yet low-priced, integrated circuits (ICs) that are available today. It is the continuation of a process which began in the mid-1960s when the first ICs began to appear. These circuits employed silicon as the active medium rather than germanium, which had been commonly used for solid-state devices up to 1960. Silicon dioxide was the key feature of this technology: it was readily produced by thermal oxidation of the silicon substrate, it was highly adherent, was insoluble in water (unlike germanium dioxide), was impenetrable by dopants and could be selectively etched with hydrofluoric acid (HF) to form windows for diffusion.

The level of complexity at that time was about four NAND gates on a single silicon chip, containing about twenty components, mostly diodes and resistors, at a density of about 3/mm². The devices were bipolar and diode-transistor logic (DTL) was used. Improved logic families soon came on the scene – TTL, I²L, ECL and so on – and analogue devices such as operational amplifiers. By 1970 the typical integrated circuit contained perhaps 100 components, many of them transistors at a density of about 20/mm², but now a great leap forward occurred: the introduction of integrated MOS devices. Though inherently simpler than bipolar devices, MOS devices had not been used much because of instability associated with charge accumulation in the gate insulation, which could lead to the unfortunate state of affairs of a transistor being permanently turned on regardless of the potential applied to the gate. Once the gate insulation had been made impervious to cation diffusion, MOS ICs that worked reliably were made and a rapid expansion in component density and concomitant price reduction for better performance was possible.

The solution of the MOS instability problem brought about the introduction of LSI (Large Scale Integration), especially for memory chips, with perhaps 5000 components at a density of 200/mm². By 1975 a further ten-fold increase in component count had been achieved and VLSI began. But the rate of increase in components per chip began to slow, possibly because the limits of optical lithography (masking and etching) had been approached with conductor widths as small as 2 μm. Further improvements were possible only with slower, more expensive, electron-beam lithography and ion implantation instead of diffusion. In spite of this, component densities have reached about 10 000/mm^2 – 10^6 per chip – and semiconductor RAM (random access memory) costs about 0.001 p/bit in 1989. Further improvement in speed and component density – by at least a factor of ten – still using silicon, is certain.

5.1 A Quick Overview of the IC Production Process

The purpose of any IC production process is to produce reliable circuits whose components are formed in a single piece of silicon (a 'chip'). These components are usually transistors and diodes, which are easy to make by the planar process shown in figure 5.1. The planar process was patented by the Fairchild Semiconductor Co. about 1965 and is so called because all the processing was carried out on a flat wafer surface. Capacitors and especially inductors are difficult to make by this method – or any other IC process – other than small values such as 100 pF or 0.1 μH, though low-power resistors are possible. Figure 5.1 is merely a rough sketch of a few of the steps in making an npn bipolar transistor. The vertical scale is much exaggerated and the silicon surface is shown flat throughout, though it is actually stepped because of the selective oxidation occurring during the various diffusions.

Figure 5.2 is a simplified IC-process flow-diagram. Considerably more space would be required to show all the stages in detail: roughly a hundred steps are involved altogether, and unless the yield at each step is very high, few circuits will be working at the end. Great care and attention are necessary to keep process yields up. Nevertheless circuit yields at the circuit-test stage are normally above 50 per cent. As the circuits become more complex and the area of each chip increases, yields will be reduced unless improvements in processing are made.

5.2 Crystal Growth and Wafer Production

Before a pure, single crystal of silicon can be grown, polycrystalline, electronic grade silicon (EGS) must first be made. Silicon comes from fairly pure quartz sand, which must first be reduced with carbon (for example, coke):

$$SiO_2 \text{ (s)} + 2C \text{ (s)} \rightarrow Si \text{ (s)} + 2CO \text{ (g)} \uparrow$$

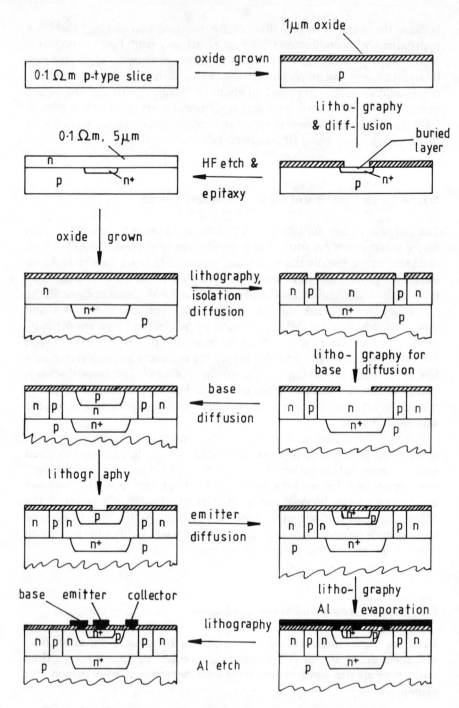

Figure 5.1 *Cross-section of a silicon slice going through the planar process (exaggerated vertical scale)*

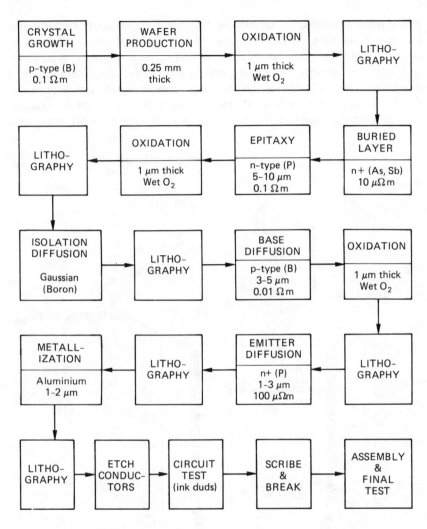

Figure 5.2 *Flow diagram for IC production*

This silicon is not very pure, but by turning it into a volatile compound which can be distilled and redistilled, the silicon can be very highly purified, for example

$$Si\ (s) + 3HCl\ (g) \rightarrow SiHCl_3\ (l) + H_2\ (g) \uparrow$$

Trichlorosilane $(SiHCl_3)$ is readily distilled to very high purity and can be reduced to form polycrystalline lumps of EGS:

$$2SiHCl_3\ (l) + 2H_2\ (g) \rightarrow 2Si\ (s) + 6HCl\ (g) \uparrow$$

Polycrystalline silicon is unsuitable for device fabrication because p–n junction formation cannot be controlled because of rapid diffusion of dopants down the grain boundaries present in polycrystalline aggregates, so it must be converted into single crystal form. The most commonly used machine for this is known as a Czochralski grower, illustrated diagrammatically in figure 5.3. Silicon is much more refractory than germanium (which is why germanium was first used in solid-state devices), melting at

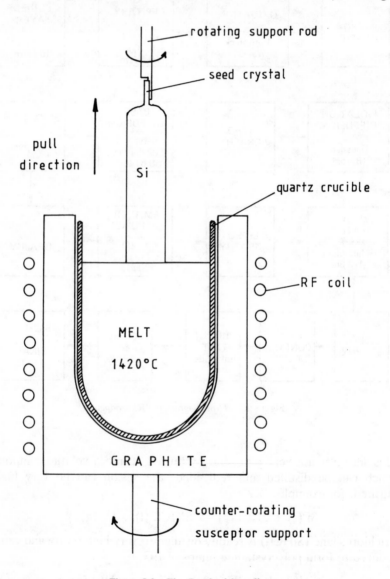

Figure 5.3 *The Czochralski puller*

1410°C (compared with 937°C for Ge) – a temperature high enough to cause considerable contamination problems; but these are almost completely overcome by wise choice of materials and design.

The silicon charge is placed in a silica crucible of very high purity, surrounded by graphite also of the utmost purity (usually nuclear-reactor grade). The graphite can be heated by radio-frequency coils (eddy current heating) or by direct resistance heating: either way, care is taken that only the graphite, the crucible and the charge get very hot. The atmosphere is usually pure argon which cannot react with anything and which helps to suppress evaporation. Carbon and oxygen are the only major impurities inadvertently introduced by the process of crystal growth. If no dopants were added the resistivity of the crystal might be about 10 Ωm, but typically what is required is p-type silicon of 0.1 Ωm resistivity, in which case boron is added to the melt in lumps of silicon heavily doped with boron.

A small, dislocation-free, seed crystal of the desired orientation is lowered into contact with the molten silicon surface and carefully withdrawn to form a cylindrical single-crystal rod up to 150 mm in diameter and 2 m long. At growth rates of 2 mm/minute, this process can take 24 hours. Both crystal and melt are rotated in opposite senses to reduce the effects of thermal and dopant-density fluctuations.

The single crystal ingot is usually ground to some exact diameter such as 127 mm and one or two flats are then ground along the length to provide an orientation mark for subsequent photolithographic operations, and to identify the conductivity type and orientation (the growth direction is usually either <100> or <111>, though <100> is preferred for VLSI). The ingot is then tested for resistivity at various places by four-point probe and cut into 0.5 mm thick slices with a circular saw. The saw blade is made from 0.3 mm thick stainless steel impregnated at its cutting surface with diamond dust. More than a third of the silicon is thus lost as dust. The sawing operation is critical since the slices must be as free from damage, as flat and as parallel as can possibly be. Each slice takes 4–5 minutes, so the whole of a 3 m ingot can take several days to process.

The rough-cut slices must next be ground flat with alumina grit (a process known as lapping) so that the thickness is uniform to a few μm. They may then be chemically etched to remove edge damage from sawing and grinding, and finally they are given a very high polish on one side by using, for example, a slurry of silica in sodium hydroxide. The lapping, etching and polishing stages remove about 100 μm of silicon so that the final slice is about 400 μm thick, and half the original ingot is lost as dust.

Though some devices may use the slices as produced from the steps so far detailed, most VLSI devices make very little use of the silicon slice itself, which is merely a mechanical support (or *substrate*) for the active silicon parts produced by epitaxy.

5.2.1 Segregation

Because the equilibrium concentrations of impurities are different in solid and melt, their concentrations must vary along the length of the pulled crystal. The phase diagram for silicon with a tiny amount of impurity takes on the very simple form of figure 5.4. From this we can see that cooling a silicon melt containing an initial concentration of impurity, C_0, causes the solid separating out at the liquidus to be lower in impurity concentration (C_s) than the melt (C_0). As cooling progresses the impurity concentration in the solid moves along the solidus, while that in the melt moves along the liquidus. The melt is left richer in impurity concentration as crystallization progresses. The segregation coefficient, k, is the ratio of the impurity concentration in the solid to that in the melt. It can be shown that

$$C_s = kC_0(1 - s)^{k-1}$$

where C_s is the concentration in the solid, s is the fraction of the melt that has solidified and C_0 is the initial concentration in the melt. Figure 5.5 shows the variation in C_s with k and s. Because its segregation coefficient is near unity (0.8), boron is the preferred dopant in the Czochralski process. See also problem 5.1.

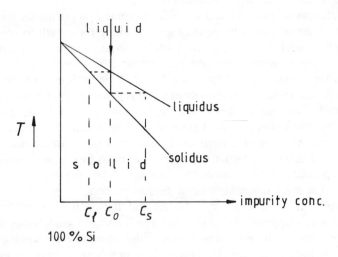

Figure 5.4 *The Si-impurity phase diagram at the Si-rich end*

5.3 Epitaxy

Epitaxy is the production of one ordered crystal structure on top of another. If the two crystal structures are the same the process is called

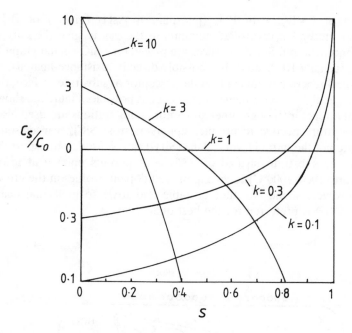

Figure 5.5   *The variation of impurity concentration in a solid with the fraction of melt solidified*

homoepitaxy, and if they are different it is called *heteroepitaxy* (but if the structures are too different, epitaxy is impossible). We are mostly concerned with homoepitaxy – silicon on silicon. What is the point? Why not form devices in the slice without epitaxy? The answer is: because we can control the electrical properties better by epitaxy than by diffusion or ion implantation – the dopant concentration is very uniform throughout an epi-layer.

Originally epitaxy was used because the collector series resistance became high in high breakdown-voltage, bipolar transistors (high V_b means high resistivity). It was lowered by the buried layer technique illustrated in figure 5.1. The p-type substrate is diffused with a high concentration of arsenic or antimony (a phosphorus buried layer would out-diffuse too much during subsequent diffusion operations) to form the n+ buried layer. A higher resistivity n-type epitaxial layer from 5 to 10 μm thick is then deposited on top and subsequent diffusions of boron and phosphorus form the base and emitter. The buried layer greatly reduces the resistance in the collector circuit. This can now be achieved by ion implantation, rather than diffusion.

The original method for achieving silicon homoepitaxy was to reduce $SiCl_4$ or $SiHCl_3$ with H_2 at about 1200°C (known as chemical vapour deposition or CVD), with PH_3 (phosphine), AsH_3 (arsine) or B_2H_6 (diborane) present in controlled amounts in the gas to give the correct dopant concentration. Substrate slices are placed on a SiC-coated graphite susceptor (usually RF-heated, but possibly directly resistance-heated), so that chemical reaction occurs only at the susceptor or substrate surface, the rest of the gas-tight reactor being kept as cool as possible. Figure 5.6 shows a horizontal reactor, though other susceptor configurations are used also. The temperature needed to produce epi-layers from $SiCl_4$, causes some out-diffusion of the buried layer. Silane (SiH_4) can be used to transport the silicon to the substrate instead of $SiCl_4$ as it permits epitaxy at lower temperatures (900–1000°C) to give improved dopant profiles in the grown layer. However, silane is rather unstable and toxic so molecular-beam epitaxy (MBE) is preferred for the best quality layers.

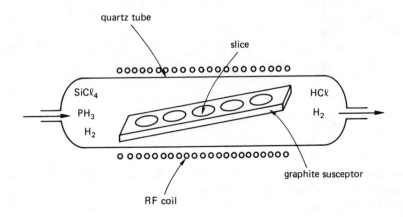

Figure 5.6 *A horizontal epitaxial reactor*

MBE essentially consists of a vacuum chamber capable of sustaining 10^{-10} torr (1 torr $\equiv$ 1 mm Hg $\equiv$ 133 Pa) – ultra-high vacuum – in which silicon (and separately, dopant) is evaporated onto the substrate. The substrate temperature controls the maximum growth rate and varies from 400°C upwards. Only very low growth rates (0.01 μm/minute) are attainable at the lowest temperature, but 0.3 μm/min can be reached at 800°C. Silicon is too refractory to permit evaporation at reasonably low temperatures so an electron beam is focused onto the EGS source, which causes local heating and evaporation. The evaporated atoms travel in straight lines to the substrate, where they condense and form the epi-layer. The thickness is monitored by a quartz crystal oscillator, in which the frequency-controlling quartz crystal is exposed to the evaporated beam of silicon atoms that condenses onto the quartz and reduces the frequency of oscillation. The frequency shift is linear with silicon thickness for thin layers. In VLSI, epitaxial layers are typically 5 μm thick, with thickness and resistivity variations over a slice of ± 10 per cent.

5.3.1 *The Evaluation of Epitaxial Layers*

The main requirements of an epitaxial layer are that it should

(1) be crystallographically perfect
(2) be of the correct and uniform thickness
(3) be of the correct and uniform resistivity
(4) permit accurate mask alignment in later processing.

The first may be checked by etching with Sirtl etch (an HF/HNO_3 mixture) and counting dislocation and stacking fault etch-pits. The epilayer is inevitably less perfect than the substrate.

The second is checked by infra-red reflectance–interference. High purity silicon is transparent to infra-red radiation, while heavily doped silicon reflects it. If an n+ buried layer is present then an infra-red beam will be reflected from the substrate/epi-layer interface and interfere with the surface-reflected beam as in figure 5.7. The interference pattern has maxima and minima whose wavelengths are given by

$$p\lambda = 2d\sqrt{(n^2 - \sin^2\vartheta)} \qquad (5.1)$$

where p is the order of the extremum (1, $1\frac{1}{2}$, 2, $2\frac{1}{2}$ etc. – half orders are minima), n the refractive index of silicon (3.42), ϑ the angle of incidence of the beam and d the thickness of the layer.

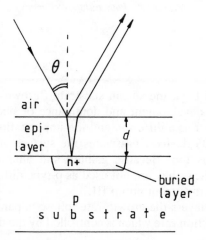

Figure 5.7 *The reflection of infra-red radiation from a buried layer*

Resistivity can be checked with the four-point probe as described before, but there are other methods such as the capacitance of a reverse-biased diode formed in the epi-layer. The capacitance of the diode is related to the dopant concentration in the layer as in equation (4.26).

The fourth requirement presents a difficulty. Figure 5.8 shows a cross-section of part of a slice before and after epitaxy. The buried layer diffusion process has resulted in about 50 nm more silicon being removed from over the n+ area of the substrate. This provides a means of mask alignment during subsequent photolithography, but the epitaxial layer has caused *pattern shift*, whereby the step in the silicon surface is shifted sideways. Pattern shift is a function of substrate orientation and can be minimized by cutting <111> slices 3° off the exact orientation towards the nearest <110> direction and by cutting <100> slices exactly on orientation. Pattern washout – disappearance of the pattern in places – also happens with slices oriented exactly on a {111} plane.

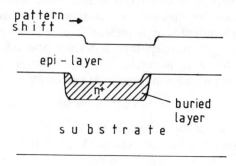

Figure 5.8 *Pattern shift after epitaxy*

5.4 Oxidation

The oxide formed from the silicon substrate ('grown' oxide) has been a vital factor in the development and dominance of silicon IC technology. Not only does it act as a diffusion and ion-implantation mask, but it also forms part of MOS devices, passivates the surface and may be used to isolate components too. Because gallium and aluminium diffuse fairly rapidly through silica, they are not used as p-type diffusants, boron being the sole representative from group III.

For thick oxide layers the growth rate follows a parabolic law, which is expected for a reaction rate which is controlled by the diffusion of reactant through the oxide:

$$x^2 = \beta t$$

where x is the oxide thickness, t the time and β the parabolic rate constant which depends on pressure, temperature and reactant species.

When the oxide layer is thin, the rate law is different, as the limiting step is the reaction at the Si/SiO_2 interface:

$$x = \alpha(t + \tau)$$

where α is the linear rate constant and τ is a constant which accounts for the initial thickness of oxide. Both α and τ depend on pressure, temperature and reactant. Table 5.1 lists some values for α, β and τ. The transition from linear to parabolic growth rate occurs at different thicknesses of oxide for different temperatures and different reactants. Substrate orientation also plays a part: α for <111> is about two-thirds bigger than α for <100>. (See problem 5.3 also.)

Table 5.1 Rate Constants for Thermal Oxide Growth on <111> Silicon

Temp. (°C)	α ($\mu m/h$)		β ($[\mu m]^2/h$)		τ (h)	
	D	W	D	W	D	W
900	0.016	0.33	0.004	0.18	1.40	0
1000	0.071	1.27	0.012	0.29	0.37	0
1100	0.30	4.64	0.027	0.51	0.076	0
1200	1.12	14.4	0.045	0.72	0.027	0

W = Wet oxygen at one atmosphere.
D = Dry oxygen at one atmosphere.

The usual oxidants employed are dry oxygen and wet oxygen. The presence of water increases both rate constants by a factor of 15–20, presumably because water is a smaller molecule than oxygen, and also reacts faster with silicon; however, wet oxide is generally of lower quality than dry-grown oxide. When masking oxide is required it must be about 0.5 μm thick, which precludes dry-grown oxide as it takes too long to form. However, low-temperature, dry-grown oxide is suitable for MOS gate oxide (only 5–50 nm thick).

Now that oxide isolation is preferred to junction isolation for VLSI devices, the limitations of these two methods' growth rates has meant greater interest in plasma oxidation, which involves the ionization of oxygen by high-frequency discharge and collecting the ionized oxygen on electrically biased silicon substrates. Growth rates are high at lower temperatures than thermal oxidation uses. CVD can also be used to deposit oxide, for example by the oxidation of silane.

5.5 Dielectric and Polysilicon Deposition

The chief uses of these deposited (usually by CVD) films in VLSI technology may be summarized:

(1) Silica heavily doped with phosphorus (usually called P-glass) is used to stabilize MOS devices against sodium ion migration and to round off sharp corners before metallization.
(2) Silicon nitride (Si_3N_4) is used as a passivating agent to stop ions diffusing towards the silicon surface, as a scratch protector over the metallization, as an oxidation mask and as a gate dielectric. Sometimes silicon oxynitride films of general formula SiO_xN_y are used instead, though these are more easily etched.
(3) Polycrystalline silicon is used as the gate electrode in MOS devices and for metallization (that is, to form the conducting paths linking devices within the IC).

The source of the silicon used in all of these materials is SiH_4, which has the great advantage of reacting at low temperatures, preventing dopant diffusion and aluminium evaporation, melting or migration – these are of much greater importance in VLSI devices, where the diffusion depths and conductor cross-sections are so small. P-glass is made from a mixture of silane, phosphine and oxygen at 500°C:

$$SiH_4(g) + 2xPH_3(g) + (2 + 4x)O_2(g)$$
$$\rightarrow SiO_2.xP_2O_5 (s) \downarrow (\text{P-glass}) + (2 + 3x)H_2O(g)$$

x is ≈ 0.05–0.08. Deposition rates are around 1 μm/hour. P-glass flows fairly well at 1100°C and sharp steps can be rounded off in 20 minutes. Sharp steps cause problems during aluminium evaporation and dielectric or polysilicon deposition, as shown in figure 5.9.

Silicon nitride is an excellent barrier to sodium and other cations which can destabilize MOS devices. It may be deposited from a silane/ammonia mixture at about 800°C:

$$3SiH_4 (g) + 4NH_3 (g) \rightarrow Si_3N_4(s) \downarrow + 12H_2 (g)$$

The film usually contains hydrogen and is more or less glassy. It is not easily etched in HF, less so as the deposition temperature is raised and the degree of crystallinity increases. The usual deposition rate is about 10 nm/minute and film thickness is controllable to 10 per cent down to 5 nm. When used as a scratch protection over the metallization, the deposition temperature has to be very low (especially when aluminium is used) – below 300°C. This can be achieved by lowering the pressure in the deposition chamber or by using plasma deposition. Plasma deposition

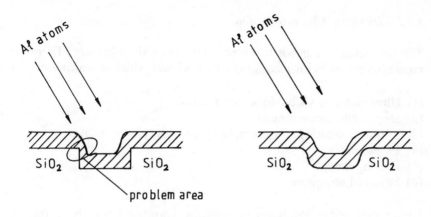

Figure 5.9 *Al evaporation onto sharp and rounded steps*

leads to amorphous, more readily etched films of lower breakdown voltage.

The silane/ammonia ratio is important for controlling the amount of silicon in the sample. Higher SiH_4/NH_3 ratios lead to films with more silicon present (not as discrete particles), which are lower in resistivity and breakdown voltage and higher in refractive index than Si_3N_4.

Polysilicon has found some favour as a conductor in VLSI because it is more refractory than aluminium, less subject to electromigration (movement of atoms due to electric fields) or corrosion and forms ohmic contacts with silicon readily. The preferred method of deposition is the pyrolysis of silane:

$$SiH_4 \text{ (g)} \rightarrow Si \text{ (s)} \downarrow + 2H_2 \text{ (g)}$$

This is done at 600°C at atmospheric pressure, but can be done at lower temperatures and reduced pressures. The deposition rate is about 20 nm/minute.

To be of use, the polysilicon must be heavily doped to reduce the resistivity to as low a value as possible. Lowering the resistivity is important for conductors, not only because of the obvious problems of voltage drop and power dissipation, but also because signal propagation is slowed by the time constant of the rails, which act as one plate of a capacitor (the substate is the other). See problem 5.4. Doping can be done by phosphine or diborane *in situ*, or by ion implantation or diffusion. The latter must be done at higher temperatures, but gives lower resistivity. Typical resistivities are 5–50 $\mu\Omega$m (for instance, $\rho_{Al} = 0.03$ $\mu\Omega$m). This is becoming a problem in high speed devices. The TCR is 0.001/K, compared with 0.004/K for aluminium.

5.5.1 Dielectric Characterization

The first thing one needs to know is how thick the films are. There are numerous methods for measuring film thickness, chief of which are:

(1) Ultra-violet or visible-light interference
(2) Step-height measurement
(3) *In situ* monitoring, for example by quartz crystal oscillator
(4) Ellipsometry
(5) Etching time
(6) Infra-red absoption.

Ultra-violet/visible interference requires knowledge of the refractive index, but if the film is deposited under standard conditions this is known. The method is non-destructive and fairly rapid. Its resolution is limited by the smallest wavelength used, for example using equation (5.1) and $n = 2$ for silicon nitride with $\vartheta = 30°$ we find

$$p\lambda = 2d\sqrt{(2^2 - 0.5^2)}$$

So if the first maximum occurs at 200 nm, $d \approx 50$ nm. This is the thinnest film that can be measured by ultra-violet/visible interference.

Step-height measurement is destructive, because a step must be cut (or formed by masking during depostion), but with modern mechanical devices (for example, Rank Taylor Hobson's Talystep[TM]) steps as small as 5 nm can be measured to 0.5 nm. This method can be used to calibrate the others: its advantage is its absolute accuracy.

Quartz-crystal frequency monitoring is normal for metallization, but less popular for dielectrics, but it works well and is an excellent process aid, since deposition is stopped when the film thickness is right and not by merely timing the deposition period. Its resolution is good – down to 5 nm. Though not an absolute standard it is easy to calibrate with the step-measurement method, and it is non-destructive and quick.

Ellipsometry is slow, and requires a rough knowledge of the film thickness, but it can determine the refractive index and absorption coefficient as well as accurately determining thickness on films down to 5 nm. Its accuracy is absolute.

Etching the film in buffered HF is destructive and crude, but can be used with moderate accuracy on films of known composition.

Infra-red absorption relies on the transparency of lightly doped silicon to infra-red radiation, and so cannot be used on processed slices after heavy doping, or metallization. It can be used on test pieces inserted into the reactor at convenient points. Its great advantage is that it tells the operator the chemical composition of the film and the degree of crystallinity. As a technique for thickness measurement it requires calibration with each type of film and is limited to thicknesses over 50 nm.

The degree of perfection of the film, particularly physical integrity, is hard to assess. On simple, conducting substrates it is possible to count pinholes and other places where the film and substrate do not cohere by dropping non-deionized water on the surface and passing a current through the water–substrate interface. Electrolysis of the water at the exposed substrate occurs and is detectable with a microscope by a stream of hydrogen bubbles. This method can throw light on problem areas by using a test wafer during processing. Composition and crystallinity are readily assessed by infra-red absorption.

Breakdown voltages can be measured at points on the film, but causing breakdown is destructive, so usually this is done on test pieces as a check. Silicon nitride has a very high breakdown voltage when it is stoichiometric and crystalline – up to 10 GV/m (or 10 kV/μm, to flout the SI convention), but the presence of excess silicon lowers this. Silicon dioxide has a high breakdown voltage when thermally grown in dry oxygen (1–2 kV/μm), but deposited films have lower breakdown voltages – about $\frac{1}{2}$ kV/μm.

5.6 Diffusion

Diffusion has been the traditional way to form p–n junctions in silicon. Only boron is used for p-type material, but a choice from phosphorus, arsenic or antimony is made for n-type. There are two diffusion profiles of interest

(1) Complementary Error Function, or erfc
(2) Gaussian.

5.6.1 Erfc Diffusion

When the source concentration at the silicon surface is constant, for example when there is replenishment from a flow of gas containing the dopant species, the diffusant concentration is given by

$$N(x,t) = N_s \text{erfc}(x/2\sqrt{[Dt]}) \tag{5.2}$$

where $N(x,t)$ is the number of atoms/m^3 at a distance x below the surface at time t; N_s is the constant surface concentration in atoms/m^3 and D is the diffusivity of the dopant. Erfc is given by

$$\text{erfc}(x) = (2/\sqrt{\pi})\int_x^\infty \exp(-t^2)\mathrm{d}t$$

A normalized plot of (5.2) is made in figure 5.10, in which $\log_{10} n$ ($n \equiv N(x,t)/N_s$) is plotted against $x/2\sqrt{(Dt)}$ ($\mathcal{X}$), from which one finds that the value of $\mathcal{X}$ remains virtually constant at 2.8 for $n < 10^{-4}$. Now the p–n junction depth will be at the point where $N(x,t) = N_0$, where N_0 is the substrate dopant concentration. For this reason, the reduced junction depth is taken as 2.8 and so

$$x_j = 5.6\sqrt{(Dt)} \tag{5.3}$$

The junction depth (x_j) depends greatly on the value of D, which in turn depends greatly on temperature, since

$$D = D_0 \exp(- E_D/k_B T) \tag{5.4}$$

where D_0 is known as the frequency factor (it is related to the frequency of the atomic vibrations in the solid) and E_D is the diffusional activation energy. Some values for D_0 and E_D for various dopants in silicon are given in table 5.2. The activation energy is comparable with the energy of vacancy formation in silicon, which points to the diffusional mechanism – interstitial diffusants would have activation energies nearer 1 eV. As a result of the high value of E_D, dopant diffusion in silicon is very temperature-dependent: see problem 5.3.

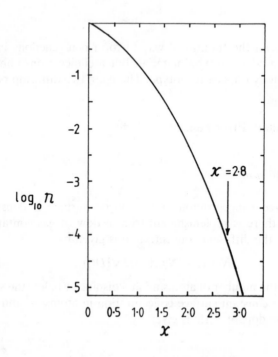

Figure 5.10 *Log_{10} (dopant concentration) against depth (an erfc plot)*

Table 5.2 Diffusional Data for Dopants in Silicon

Dopant	B	Al	P	As	Sb
D_0	0.8	8	4	21	0.2
E_D	3.5	3.5	3.7	4.1	3.7

Note: D_0 is in units of 10^{-4} m²/s, E_D is in eV.

For instance, suppose the surface concentration of phosphorus is 10^{24} atoms/m³, and we wish to form a junction at a depth of 2 μm in the base region of an npn transistor, which has a boron concentration of 10^{20} atoms/m³. The collector has an n+ buried layer doped with arsenic. If the diffusion time is 1 hour, what temperature must be employed and how far will the buried layer diffuse?

From (5.3) we find $D = 3.3 \times 10^{-17}$ m²/s, and from table 5.1 and equation (5.4) we have

$$3.3 \times 10^{-17} = 4 \times 10^{-3} \exp \left[\frac{-3. \times 1.6 \times 10^{-19}}{1.38 \times 10^{-23} \, T} \right]$$

from which we find $T = 1050°C$. At this temperature the diffusivity of arsenic is 5.2×10^{-19} m²/s, so equation (5.3) indicates that it diffuses about 0.24 μm. Under similar conditions antimony would have diffused 0.14 μm. Arsenic and antimony atoms therefore diffuse only about 10 per cent as far as phosphorus atoms, which is why they are preferred for n+ buried layers.

5.6.2 Gaussian Diffusion

If the source of dopant at the surface of the silicon slice is not replenished but is depleted by diffusion into the substrate alone, then the dopant concentration is given by

$$N(x,t) = N_s \exp(-x^2/4Dt) \tag{5.5}$$

where N_s is now time-dependent:

$$N_s = S/\sqrt{(\pi Dt)} \tag{5.6}$$

S is the original surface concentration in atoms/m². Boron is sometimes diffused into silicon in this way: boron-doped oxide is predeposited onto the slice and then boron is diffused into the substrate at a higher temperature – a process known as *driving in*. Gaussian diffusion is useful for keeping down the surface concentration of dopant.

The diffusion of dopant following ion implantation is also Gaussian. For instance, boron is implanted into silicon at a concentration of 10^{19} atoms/m^2 and a depth of 0.2 μm. The silicon is n-type with 10^{22} atoms/m^3. If the slice is held at 1050°C for an hour, where will the junction be?

The boron diffusivity is 3.8×10^{-18} m^2/s. N_s is 4.8×10^{25}/m^3 from (5.6), so (5.5) gives

$$10^{22} = 4.8 \times 10^{25} \exp(- x_j^2/4 \times 3.8 \times 10^{-18} \times 1323)$$

Solving, we find the junction depth to be 0.4 μm from the implanted layer, or 0.6 μm from the surface.

5.6.3 Diffusion Profile Measurement

We have already seen how diode *C–V* measurements can give a measure of the dopant concentration in the depletion region. Another technique which is easily used is simply to measure the sheet resistance of the diffused region and assume a diffusion profile which will give junction depth if the dopant concentration in the substrate is known. The sheet resistance is given by

$$R_s = Rw/l = \rho/d$$

where R is the resistance of a sheet of material of width, w, length, l, and thickness, d. If, for example, the resistance of a piece of material 4 mm long and 1 mm wide were 4 Ω, its sheet resistance would be 1Ω/square, and were it 10 μm thick, its resistivity would be 10^{-5} Ωm.

A technique much used for resistivity, and hence dopant, profiles is the spreading resistance probe, which is just two fine-pointed metal probes placed on the surface of the slice, the resistance between which is

$$R = \rho/2a$$

where ρ is the resistivity of the silicon and a is the probe radius. Because the probe radius can be made very small, the technique is capable of high spatial resolution, especially when coupled with slice bevelling as shown in figure 5.11. If the bevel angle is 3°, then the spatial resolution is amplified by cot 3°, or about 20 times. By this technique with accurate positioning on the bevel, dopant profiles can be obtained with a resolution of 0.02 μm.

Junction depth may also be measured by a bevel-and-stain technique. The slice is bevelled to amplify the vertical dimension and stained with a solution of copper sulphate in a mixture of hydrofluoric and acetic acids. The copper deposits preferentially on the n-type material, which is stained dark. If a microscope slide is placed on the surface of the slice as in figure 5.11, and illuminated with sodium light, the interference fringes from the

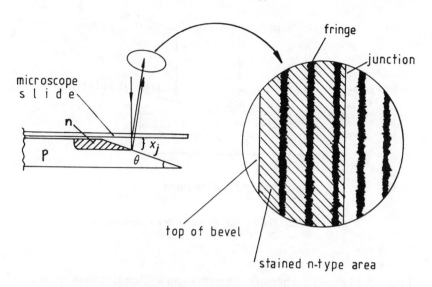

Figure 5.11 *The bevel-and-strain technique and interference fringes*

top of the bevel to the junction may be counted. The junction depth is then calculated (assuming normally incident light) from

$$2x_j = p\lambda_{Na}$$

p is the number of fringes, λ_{Na} is 590 nm. Since it is possible to measure to a quarter of a fringe, the accuracy is fairly good.

5.7 Ion Implantation

If ions of arsenic, phosphorus or boron are accelerated to sufficiently high energies they are able to penetrate the surface of a silicon slice to depths of up to 1 μm. The depth is determined by the kinetic energy of the ion, 1 μm being attained with about 0.5 MeV. Areas which are to be implanted are defined by windows in photoresist, or a contact mask, or by oxide masking when the energy is not too high. Ion implantation causes a fair amount of damage to the silicon, which is removed by annealing afterwards. The advantage of ion implantation is that sharp junctions and precise dopant concentration profiles are obtainable within very small dimensions. It can also be used in conjunction with MOS gate-contact polysilicon, which acts as a mask, to produce self-aligned source and drain, as in figure 5.12. The disadvantages are its high capital cost and slow processing rate, since only one wafer, or part of one, is exposed at a time.

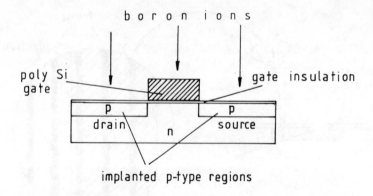

Figure 5.12 *The self-aligned gate with ion implantation*

Figure 5.13 shows a schematic diagram of an ion implantation apparatus, whose essential parts are an ion source, an analyzer to select the right ions, an accelerating potential, deflection plates to aim the ion beam, a target and slice-changing mechanism and a charge integrator to monitor the ion dosage. The whole apparatus operates in a vacuum to prevent ion scattering.

The usual sources of ions are gases such as PF_3, BF_3, and AsF_5, though the hydrides may also be used. Ionization causes a number of positive species to be produced such as B^+ (desired), BF^+, F_2^+ and F_3^+ (not desired), which travel through the magnetic field of the analyzer. The analyzer consists of a bent tube and a magnetic field arranged so that the ions are turned in the direction of the bend by the Lorentz force. The Lorentz force is counteracted by the centripetal force as the ion describes a circular arc, which must have the same radius as the bend for the ion to emerge:

$$qvB = Mv^2/R$$

or

$$B = Mv/qR$$

where B is the magnetic induction (or $\mu_0 H$) in Tesla, M is the mass of the ion, q is its charge, v its velocity and R is the radius of the bend. Now the ion's kinetic energy comes from the accelerating potential of the ionizer, V:

$$\tfrac{1}{2}Mv^2 = qV$$

So for the arc radius to equal the tube radius, R, B must satisfy

$$B = Mv/Rq = \sqrt{(2VM/qR^2)}$$

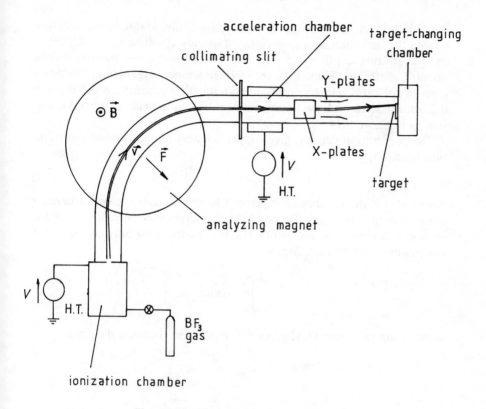

Figure 5.13 *The ion-implantation apparatus*

Adjustment of the magnetic field to the desired value will thus select the right species. A typical value for B might be 0.5 T, which is readily obtained from an iron-poled magnet.

The ions passing the analyzer are accelerated to the right energy and the ion beam is scanned over the area of an aperture exposing the target. On striking the target the ions give up their positive charge, which is neutralized by electrons from the target holder. These electrons are supplied via an integrator, which measures the total charge delivered. The dose rate is defined by

$$S = Q/qA = (1/qA) \int I dt$$

where S is the number of atoms delivered to the target per unit area and A is the aperture area. Suppose the integrator current (the same as the beam current) is 1 μA and the wafer is exposed for 1 minute through an aperture of area 10^{-4} m^2, then for a singly ionized species, S will be 3.7×10^{18}/m^2. If the penetration depth is 0.5 μm, then the dopant concentration will be about 7×10^{24}/m^3. Some ions are stopped sooner, others later, so that a layer of dopant is actually produced underneath the surface, which is about 0.2 μm thick, though the implantation profile is actually a skewed Gaussian as in figure 5.14.

The penetration depth is given approximately by the Gaussian distribution function:

$$n(x) = n_0 \exp(- [x - x_0]^2/2\sigma^2) \tag{5.7}$$

where $n(x)$ is the number of implanted atoms per unit volume at depth x below the surface, x_0 is the depth of maximum concentration, σ is the standard deviation of the depth (known as the *straggle*) and n_0 is the maximum concentration. Since

$$S = \int_0^\infty n(x)\mathrm{d}x \tag{5.8}$$

substituting (5.7) into (5.8) gives $S = n_0\sigma\sqrt{(2\pi)}$, because if $x_0 > 3\sigma$

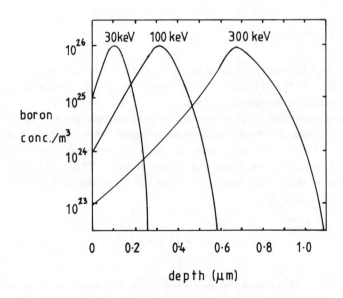

Figure 5.14   *Implantation profiles for B in Si*

then

$$\int_0^\infty \exp(- [x - x_0]^2/2\sigma^2) \approx \sigma\sqrt{(2\pi)}$$

so that

$$n_0 = S/\sigma\sqrt{(2\pi)}$$

Hence

$$n = (S/\sigma\sqrt{[2\pi]})\exp(- [x - x_0]^2/2\sigma^2) \qquad (5.9)$$

The straggle depends on the energy of the ions, the ionic species and the dose rate. The penetration depth depends on the energy and species. Boron is not readily stopped by silicon nuclei, but mainly by silicon electrons, so its penetration is greater than phosphorus or arsenic, which are stopped by silicon nuclei at moderate energies. Table 5.3 gives some ion ranges and straggles for various dopants in silicon. Figure 5.14 shows some typical implantation profiles for boron in silicon.

Table 5.3 Ion Implantation Parameters for Silicon

Ion:		B^+		P^+		As^+	
Parameter:		x_0	σ	x_0	σ	x_0	σ
E N E R G Y	k e V	10	0.04	0.02	0.015	0.008	0.011	0.004
		30	0.11	0.04	0.04	0.02	0.023	0.009
		100	0.31	0.07	0.14	0.05	0.07	0.03
		300	0.66	0.11	0.41	0.11	0.19	0.07

Note: All distances in μm.

5.7.1 Annealing Implanted Layers

One finds that after ion implantation, the dose delivered does not correlate well with the majority carrier concentration and generally at any place in the implanted layer $N_A \gg p$ or $N_D \gg n$. That is, the donor dopants do not appear to have given up their electrons to the conduction band, nor the acceptor atoms to have accepted electrons from the valence band. This is because the silicon is heavily damaged by the implantation process, causing the charge carriers from the dopants to be trapped, for example at dislocations. When high doses are used (as in emitters, buried layers and low-value resistors) so many silicon atoms are displaced from their lattice sites that the structure is amorphous, which has two advantages: firstly, it produces a more controlled implantation depth and secondly, the amorphous regions readily undergo recrystallization during which trapping sites are removed. Whether amorphous or not, the implanted layers must be annealed by heating or by using a laser. A pulsed, high-power laser can anneal the damage without significant diffusion, which can be vital for devices with very shallow junctions. Phosphorus and arsenic can be annealed by heating to about 600–700°C, but boron requires 900–1000°C, possibly causing undesirable diffusion of the dopant (see problem 5.6).

5.8 Lithography

Lithography is the process by which a pattern is transferred to the surface of the wafer. This pattern defines areas on the wafer which are to be exposed to diffusants (as in the formation of base and emitter regions of transistors), or to ion beams, or to metal etchants etc. The pattern may be made in the form of an opaque mask, which covers certain areas of a sensitive medium known as a *resist*. Typically, the resist is polymerized by exposure to electromagnetic radiation (which requires a mask) or an electron beam (which needs no mask), and this polymerized resist is insoluble in a chemical solution that removes the soluble unpolymerized resist. Exposure of the wafer to a suitable etchant then removes material not covered by resist (it is because the resist resists etching that it derives its name).

Because the process of lithography determines the size and shape of the devices made in the silicon wafer it is of great importance; in fact, the set of masks for the entire process of production of a VLSI chip can be thought of as the process definition. In the planar process briefly outlined in section 5.1 the lithography was entirely achieved by optical means, which lead to line widths of about 5 μm for conductors, since the wavelength of light is about 0.5 μm, and diffraction limits the resolution of the optical system and mask. In VLSI, ultra-violet radiation and, increasingly, electron beams are used to expose the sensitive resists. The wavelength of ultra-violet radiation is about 200 nm, so the minimum line width or feature size is about 2 μm – perhaps less if size variation of more than 10 per cent is allowed. Electron-beam resists can be found which allow reproducible features down to 0.5 μm, a figure which will certainly be bettered in time.

Figure 5.15 shows the steps in a lithographic operation leading to the formation of a window in the thermally-grown oxide. Before this process can occur, masks have to be prepared which define the pattern ultimately to be achieved in the oxide layer. These are usually generated by computer as 'artwork' which is then photographically reduced many times till the proper size is attained. Computer-aided design (CAD) has made great

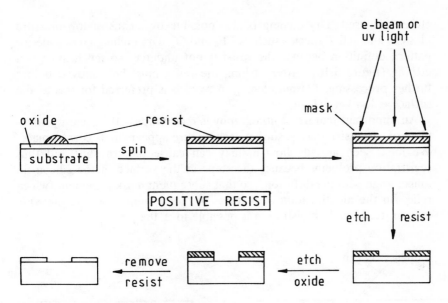

Figure 5.15 *The lithographic process for a positive resist*

strides in the preparation of masks, leading to considerable speeding up in the process of chip design and also enabling customization of chips. In the electron-beam process the circuit layout is used to generate the beam control programme automatically. The mask usually consists of a film of metal (chromium is popular as it resists scratching and corrosion) on glass. Once the mask for a process-step of a single chip has been made it can be repeated by mechanical means any number of times that may be required to cover the whole surface of a wafer. The set of such wafer masks then defines the structure of the final device.

The first step in lithography is to cover the wafer with resist, achieved by placing the wafer on a vacuum chuck, placing a drop of resist at the middle and spinning fast enough to produce a film of the required thickness (changing viscosity is a good way to change resist thickness with a fixed speed of rotation). This resist is then dried in an oven and exposed. The exposed resist is then etched in a proprietary solvent. In the case of negative resists, the polymerized material is insoluble, and while for positive resists, the exposed resist is rendered soluble. Thus, a positive resist is one which renders the mask positive, that is the mask defines the area which is eventually to be diffused or implanted, though for conductors the position is reversed. When the resist pattern has been established in this way, the resist is hardened by baking and rendered insoluble in the etchant for the oxide or metal now exposed. Silicon dioxide and silicon nitride are usually etched in HF and aluminium in phosphoric acid, if wet etches are used. Dry etching is also possible by means of low-pressure plasmas and active species such as CF_4 and O_2. Dry etching gives superior pattern definition because the resist is not undercut so much as in wet etching (figure 5.16). After etching the resist must be removed before further processing. Plasma etching in oxygen is preferred for this as the surface is left very clean.

An important feature of lithography is *registration* – the accurate alignment of successive masks, so that each feature appears in its correct place. We have seen already how epitaxy can cause pattern shift and thus registration problems because the steps in the surface of the silicon no longer align with the diffusions, so that subsequent mask alignment (which relies on the identification of features by, for example, surface steps) is inaccurate. Usually registration is possible to ± 0.2 μm.

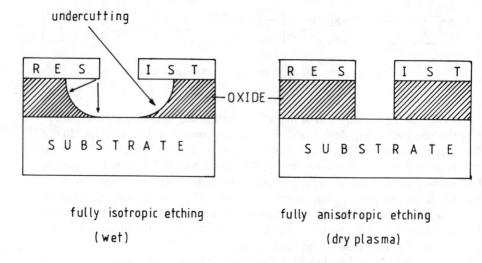

Figure 5.16   *Isotropic (wet) and anisotropic (dry) etching*

5.9 Metallization

Metallization refers to the formation of contacts and conductors on the surface of the chip. Aluminium is the preferred conductor because it is readily evaporated and sticks well to silicon dioxide, though it suffers from being rather readily scratched and corroded and is subject to electromigration. If aluminium is used as a direct contact to silicon, problems arise because of the solubility of silicon in aluminium. For this reason metal silicides have been used as they will not dissolve silicon, have low resistivity and form ohmic contacts.

Aluminium is invariably evaporated in a vacuum from a heated refractory boat made of Al_2O_3 or BN. The deposition rate is monitored by a quartz crystal oscillator, and the slices remain fairly close to room temperature. Deposition rates of 0.5 μm/minute are usual.

5.9.1 Contacts

Contacts may be rectifying or ohmic. The rectifying nature of the Al–Si contact is made use of in Schottky-barrier diodes (Schottky published the theory of metal–semiconductor contacts in 1938). The contact resistance is dependent on two things: the contact potential barrier height, ϕ_B, and the dopant concentration in the silicon. When the doping in the semiconductor is low, electrons from the metal enter the semiconductor by thermionic emission over the barrier, a process governed by the Richardson equation:

$$J_s = A^* T^2 \exp(- e\phi_B/k_B T)$$

where J_s is the reverse saturation current density and A^* is the effective Richardson constant ($\equiv 4\pi m^* e k_B^2/h^3 \approx 10^6$ A/m^2/K^2). The diode equation gives the current density for forward and reverse bias conditions:

$$J = J_s(\exp[eV/k_B T] - 1) \tag{5.10}$$

The specific contact resistance is defined by

$$R_c = (\partial V/\partial J)_{V = 0}$$

R_c is in units of Ωm^2. From equation (5.10) we find

$$R_c = (k_B/eA^* T)\exp(e\phi_B/k_B T) \tag{5.11}$$

ϕ_B is about 0.7 V for an aluminium contact to n-type silicon and about 0.6 V for an aluminium contact to p-type silicon. For n-type silicon at room temperature, (5.11) gives $R_c \approx 0.1$ Ωm^2, a very large value; see problem 5.8.

Equation (5.11) holds for low doping levels and is independent of the doping concentrations for dopant concentrations less than about 10^{24}/m^3. However, at higher doping levels the electrons can tunnel through the barrier and R_c then depends on the dopant concentration also:

$$R_c \approx 10^{-9}\exp(4.4 \times 10^{13}\phi_B/\sqrt{N}) \tag{5.12}$$

where N is the dopant concentration. Figure 5.17 shows a plot of (5.12) for various values of ϕ_B and N, from which we can see that the minimum value of R_c is about 10^{-8} Ωm^2. When $N < 10^{24}$/m^3, the specific contact resistance becomes constant at the value given by (5.11). For ohmic contacts, the silicon in contact with the metallization must be highly doped.

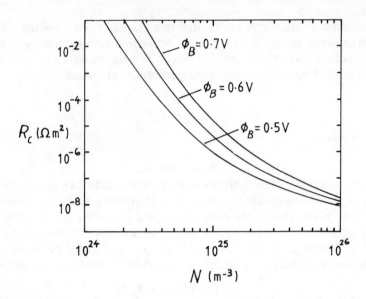

Figure 5.17 *R_c against dopant concentration for various contact potentials*

5.9.2 *Junction Spiking*

Figure 5.18 shows the phase diagram for Al–Si at the Al-rich end. There is a eutectic at 577°C, which is the upper limit to which the wafers may be heated during or after aluminium deposition. However, even below the eutectic temperature, there is some solid solubility of silicon in aluminium: at 450°C aluminium dissolves about 0.4 weight per cent of silicon. The diffusion length is $\sqrt{(Dt)}$, where D is the diffusivity for Si in Al, given by $D_0\exp(-E_D/k_BT)$, with $D_0 = 4 \times 10^{-6}$ m²/s and $E_D = 0.92$ eV, so that after an hour at 450°C, the silicon has diffused about 75 μm – laterally, as well as vertically above the contact. Consider the contact shown in figure 5.19. The aluminium rail width is 10 μm and its thickness is $\frac{1}{2}$ μm. The volume of aluminium saturated with silicon is $75 \times 10 \times \frac{1}{2} (= 375)$ (μm)³. The volume of silicon dissolved (ignoring the slight density difference between aluminium and silicon) is therefore 0.004×375 (μm)³. If the contact window is 5×5 (μm)², the depth of silicon dissolved is 0.06 μm. Unfortunately, the silicon is dissolved preferentially at certain sites to form spikes filled with aluminium, which can penetrate through to the junction in shallow device structures.

Spiking can be avoided by saturating the aluminium with about 1 weight per cent silicon during evaporation, or by using a polysilicon or silicide contact. Deposition is by co-evaporation, for example of tantalum and silicon, to form $TaSi_2$ after sintering at 1000°C. Silicide resistivities

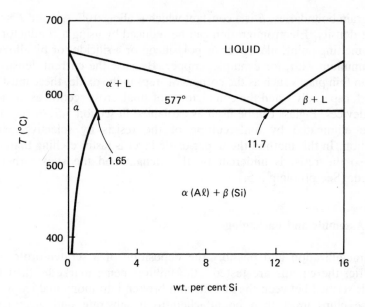

Figure 5.18 *The Al–Si phase diagram at the Al-rich end*

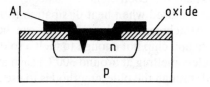

Figure 5.19 *Junction spiking by Al*

are of the order of 0.5 μΩm (compare $\rho_{Al} = 0.03$ μΩm and $\rho_{Si}(min)$ = 10 μΩm).

5.9.3 Electromigration

The impact of electrons on aluminium atoms at high current densities leads to a migration of the aluminium towards the direction of higher potential (that is, in the direction of electron flow). The mean time to failure of an aluminium conductor is given by

$$\tau = BJ^{-2}\exp(E_A/k_BT) \tag{5.13}$$

where the activation energy, $E_A \approx 0.5$ eV, suggesting that migration of atoms in the bulk is not responsible (for which $E_A \approx 2$ eV), but movement along grain boundaries. B is a constant which is about 10^{18} A^{2}s and J is the current density. Electromigration can be reduced by using a conductor of higher melting point, like silver or polysilicon or a silicide, or by alloying the aluminium with, for example, copper. Because the current density is raised in thin places such as the corners of steps in the oxide, these must be avoided, especially when the conductor is of small cross-section as in most VLSI devices. P-glass can be used as explained in section 5.5, or the step can be eliminated by undercutting of the resists by selective oxide deposition. In this method the upper oxide layer is faster etching than the lower so the resist is undercut by the etchant and the step corner is rounded. (See problem 5.5.)

5.10   Assembly and Packaging

After metallization and possibly the deposition of a scratch-protectant dielectric, the circuits are tested – the failures being marked – then the wafer is scribed between the circuits and broken into individual chips.

These chips must then be attached to a substrate and electrically connected to leads as in figure 5.20. The substrate can be ceramic, such as alumina or beryllia, or metallic, such as Kovar (an Fe–Ni alloy with a similar thermal expansion coefficient to silicon). Ceramic substrates (especially beryllia) are used when heat dissipation from the chip is likely to be a problem. Attachment to the substrate can be accomplished by heating the substrate and chip hot enough to melt a solder pre-form placed under the chip. Solders melting at around 300°C, such as Pb–Sn, Au–Sn or Au–Ge can be used, though the oldest method is to use a gold preform and 'scrub' (scrubbing just means rubbing the chip on the preform). Gold and silicon form a eutectic at 3.6 weight per cent Si, which melts at 370°C, so that when the gold preform is heated to above this temperature it begins to dissolve silicon and melts. The further dissolving of silicon raises the melting point until the alloy solidifies again. 'Scrubbing' the chip assists this process.

The electrical connections from chip to the leads must be made next. In the most common case, gold wire is fed through a metal tube where it is formed into a ball by a hydrogen flame, as in figure 5.21(a). This ball is then pressed onto the contact pad on the chip and a bond is made by applying ultrasound to the ball via the metal tube (figure 5.21(b)). The tube is then moved across to the lead-frame bonding pad (which connects to the pins that the customer sees outside the package) and there it is bonded ultrasonically (figure 5.21(c)). The wire is then severed and flamed and the process is repeated. High bonding rates are possible by this means and high reliability is achieved.

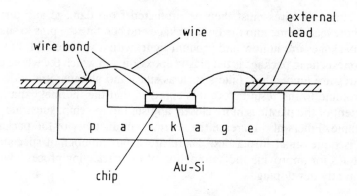

Figure 5.20　*Chip bonding to package*

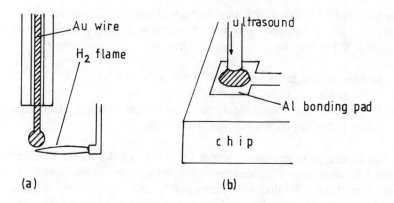

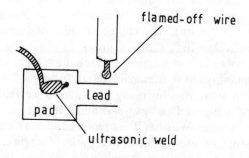

Figure 5.21　*Steps in wire bonding*

The chip and wires must then be protected from damage and dirt by sealing the lead-frame into the final package and bending the pins to shape. Some packages are hollow and vacuum-tight, with a soldered lid, but the commonest cheap package is the plastic epoxy one, in which the whole chip and wires are embedded. Fillers such as silica and alumina are added in large volume to the resin before it is cured so that the thermal expansion coefficient of the plastic is more closely adjusted to the chip, substrate and lead-frame. Fillers also increase the thermal conductivity of the package, which is quite small in plastics. There are a fair number of specialized techniques for improving the automation of the packaging process, which is constantly developing.

5.11 Beyond Silicon

For twenty-five years silicon has dominated the semiconductor industry, and one is given to wonder how much longer it will remain the preferred IC material. What are the properties required of a semiconductor?

(1) It must have a fairly large band-gap – at least 1 eV.
(2) The carrier mobilities should be as high as possible.
(3) It must be not too difficult to make single crystals which can be controllably doped both n-type and p-type.

The band-gap requirement is met by nearly all the materials listed in table 5.4, but few have higher mobility than silicon – only InP and GaAs in fact. The third condition eliminates highly refractory materials such as diamond and SiC, and a lot more not listed. We are then left with a choice of mostly III–V and II–VI compounds, and compound semiconductors are harder to use than elemental semiconductors, because of problems with dissociation and non-stoichiometry (it is difficult to make anything but n-type CdS for this reason). The choice of elemental semiconductors is woefully limited: other than those in group IV only boron (impossibly refractory), phosphorus (too reactive) and selenium (difficult to make and with very low mobility) have reasonably large band-gaps.

GaAs has been advocated for many years, and it certainly scores over silicon in having a higher band-gap and much higher electron mobility, promising faster devices. Why has it not replaced silicon? Undoubtedly the major reason is that it is more difficult to grow large single crystals of GaAs and it does not form an adherent oxide for masking purposes. These factors make it more costly to process into ICs, but not impossible – GaAs ICs have been around for almost as long as silicon ICs. But in truth the limitations to VLSI technology have little to do with the limitations of silicon used as semiconductor, but much more to do with speed and

Table 5.4   Mobilities and Band-gaps at 300 K

Material	Crystal Form	Melting Point (K)	Band-gap (eV)	Mobilities ($m^2/V/s$)	
				Electron	Hole
Diamond	A4	4300	5.4	0.18	0.14
Si	A4	1683	1.11	0.14	0.05
Ge	A4	1210	0.66	0.38	0.18
β-SiC	B3	3070	2.3	0.01	0.002
AlAs	B3	1870	2.16	0.12	0.04
AlSb	B3	1330	1.60	0.09	0.055
GaP	B3	1750	2.24	0.03	0.015
GaAs	B3	1510	1.43	0.85	0.04
InP	B3	1330	1.35	0.50	0.015
α-SiC	B4	3070	2.86	0.04	0.002
ZnS	B4	2100	3.67	0.02	0.01
ZnSe	B4	1793	2.58	0.054	0.003
CdS	B4	1748	2.42	0.04	0.002
CdSe	B4	1512	1.74	0.065	0.004
CdTe	B4	1200	1.44	0.12	0.006

Notes: *Strukturbericht* symbols:
A4 – diamond
B3 – sphalerite (zincblende)
B4 – wurtzite (zincite).
Mobilities given are highest values from Hall effect data.
Melting points are sublimation temperatures in some cases.

temperature limitations imposed by the interconnection and packaging technologies. To consider the temperature problem further: what are the limits for silicon? Roughly speaking, the highest operational temperature for a semiconductor is determined by its band-gap; silicon's limit is about 250°C, which is higher than the normal interconnections will tolerate. GaAs will probably go up to 350°C – not a huge improvement. Switching times are limited more by the interconnections than by the mobility of charge carriers in the semiconductor, and though ballistic devices (in which the charge carriers do not have to reach thermal equilibrium) may make an impact here, the interconnection problem will remain. Photonic devices are another possibility which will be considered later. It would be a brave man who forecast the demise of silicon for the foreseeable future (that is, until the 21st century).

Problems

5.1. Figure P5.1 shows a part of a hypothetical phase diagram for Si–X at the silicon-rich end. Estimate the segregation coefficient (*k*) for X in

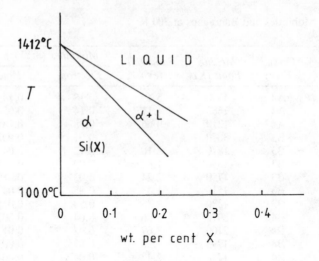

Figure P5.1 *The Si–X phase diagram at the Si-rich end*

Si from it. If $C_0 = 10^{-7}$ weight per cent and the permissible concentration variation for X in a pulled single crystal is 20 per cent, what percentage of the melt must be left?

[Ans. 0.6; 51 per cent]

5.2. Figure P5.2 shows an infra-red reflectance–interference spectrogram for an epitaxial layer on silicon. Use equation (5.1) to find the epi-layer thickness if $\vartheta = 30°$.

[Use all of the maxima and minima, take the average and standard deviation, which is an indication of the precision of the method. Phase shifts at the interface must be ignored, but normally are taken into account.]

[Ans. 15.9 μm; $\sigma_{n-1} = 0.25$ μm]

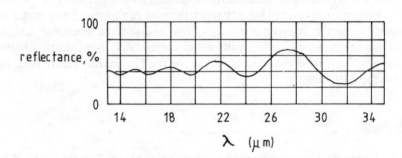

Figure P5.2 *An infra-red reflectance–interference spectrogram*

5.3. It is desired to grow, by dry oxidation of <100> silicon, a gate insulation layer 40 nm thick in 30 minutes. What is the lowest temperature at which this can be done?

[The transition from a linear to a parabolic growth law occurs roughly at $x = \beta/\alpha$.]

Suppose 1 μm of wet oxide is grown on a <111> silicon slice. What depth of silicon is consumed by this process? Windows are cut in this oxide and further oxidation occurs in wet oxygen at 1100°C for 1 hour. Find the oxide thicknesses (a) over the window and (b) over the rest of the slice. Hence find the height of the step in the silicon when the oxide is removed in HF.

[Densities: Si, 2330 kg/m^3; SiO$_2$, 2200 kg/m^3.]

[Ans. 1000°C; 0.44 μm; (a) 0.714 μm; (b) 1.229 μm; 0.214 μm]

5.4. The length of a conductor rail is l, its thickness is d and its resistivity is ρ. It passes over dielectric of permittivity ϵ and thickness b. Show that the time constant for signal propagation in the conductor is

$$\tau = RC = \epsilon\rho l^2/bd$$

[Consider the conductor, the dielectric and the substrate as a capacitor and ignore the substrate resistance. This shows that the most important parameter for propagation delays is the *length* of the conductor.]

5.5. Silicon oxynitride is deposited on a silicon slice with increasing amounts of oxygen as the layer thickens, so that its etch rate in buffered HF varies linearly from 60 nm/minute at the substrate to 100 nm/minute at the top of the layer. Show that the time in minutes taken to etch away x nm is given by

$$t = -0.025\xi\ln(1 - 0.4x/\xi)$$

where ξ is the layer thickness in nm. From this equation it is possible to find how long any particular depth has been etched, and hence how far the layer has been etched laterally. Consider a layer 1 μm thick and draw the vertical step profile.

5.6. An isolation diffusion is carried out on a 5 μm thick epi-layer doped with 10^{21} phosphorus atoms/m^3, by using boron from a depleting surface source of initial strength 10^{21} atoms/m^2. How long will the process take at 1200°C? If there is a buried layer doped with arsenic of effective surface concentration 10^{23} atoms/m^2, how far will this diffuse during the isolation drive-in? Repeat this calculation for an antimony-doped buried layer, then repeat all the calculations for a drive-in temperature of 1150°C. Comment.

[Take the buried-layer position to be the point where the donor concentration reaches 10^{22} atoms/m^3.]

[Ans. 1 h 28 m; 1 μm; 0.47 μm. 3 h 50 m; 0.9 μm; 0.45 μm]

5.7. A silicon epitaxial layer is grown using silicon tetrachloride, which is taken up from a liquid source with hydrogen. The temperature of the $SiCl_4$ bath is maintained so that the vapour pressure is 100 mm. If the hydrogen flow-rate is 10 litres/minute, what is the maximum rate of growth of the epi-layer if the heated zone in the reactor has an area of $0.1 \ m^2$?

Suppose that phosphorus atoms are incorporated into the epi-layer with the same probability as silicon atoms, so that their concentration in the solid is the same as their concentration relative to silicon in the gas. What will the concentration of phosphine have to be in the gas phase in the previous part of the question if the resistivity of the epi-layer is to be 0.1 Ωm at 20°C?

[Ans. 7 μm/minute; 1.05 ppb]

5.8. Consider a square contact region between aluminium and n-type silicon which is 10 μm along a side. What must the dopant concentration be for the contact to pass 0.1 mA at a forward voltage of 0.1 V? [Ans. $4.5 \times 10^{25}/m^3$]

6

Magnetic Phenomena

Magnetism has a very ancient history: lodestone, a naturally-occurring permanent magnet, was known as early as 800 BC. Lodestones were of vital importance to navigation for making compasses, described by Alexander Neckam, an Englishman, in about AD 1200. William Gilbert's *De Magnete*, published in 1600, began the scientific study of magnetism which languished until the rapid succession of discoveries in the 19th century, starting with Oersted's observation of the magnetic field of an electric current in 1820. In this chapter we shall briefly describe the principal kinds of magnetic ordering to be found in solids and shall then go on to consider in more detail the phenomena that are most relevant to electrical engineering

6.1 Magnetic Units

Magnetism is bedevilled by the different units used for magnetic field strength, H (SI unit, A/m) and magnetic induction, B (SI unit, T for tesla) and by the fact that most publications still use the old c.g.s. units of oersted (Oe) and gauss (G) for these respective quantities. In addition, there are two rival subsystems of magnetic units within the SI. One of these, the Sommerfeld system, says that when a magnetic field, H, is applied to a material it may be considered to give rise to a magnetic induction, B, which is related to H by

$$B = \mu_0(H + M) \tag{6.1}$$

M is the magnetization or the magnetic moment per unit volume of the material (and has also units of A/m). B, H, and M are all vector quantities. μ_0 is the permeability of a vacuum, $4\pi \times 10^{-7}$ H/m, though it is better to call it the magnetic constant. Figure 6.1 shows the three fields for a ferromagnetic material. Because $B = \mu_0 H$ in a vacuum, a magnetic field is often said to be, say, 0.5 T instead of $0.5/\mu_0$ (= 398 000) A/m. Some say this is incorrect, but it is convenient, because a good many equations then do not require a factor of μ_0, while conversion from tesla to gauss or oersteds is simple (1 T = 10 000 G, 1 G = 1 Oe). In the Kennelly system, equation (6.1) may be written in terms of the magnetic polarization, $J(\equiv \mu_0 M)$:

$$B = \mu_0 H + J \qquad (6.2)$$

J is in tesla. We shall use the Kennelly system most of the time, as fewer μ_0s are required in the formulae. Since J and M are the same quantity apart from the magnetic constant, to speak of the magnetization doing such and such is the same as saying that the magnetic polarization does so too.

In the c.g.s.–e.m.u. system, equations (6.1) and (6.2) are written

$$B = H + 4\pi M \qquad (6.3)$$

so that in a vacuum B and H are numerically equal, but in different units of gauss (G) and oersteds (Oe) respectively. The c.g.s. unit of magnetization is the gauss, but a factor of 4π is needed to convert this to magnetic induction, as shown by (6.3)

Using just magnitudes, dividing (6.1) by H gives

which can be written $$B/H = \mu_0(1 + M/H)$$

$$\mu = \mu_0(1 + \chi) = \mu_0 \mu_r$$

where μ is the (magnetic) permeability of the material, χ is its (magnetic) susceptibility and μ_r is its relative permeability, usually referred to in the magnetic literature as *the* permeability, because in the c.g.s. system $\mu_0 = 1$ and $\mu = \mu_r$. χ is dimensionless in the SI system. The permeability of a material is *not* a constant, except under specific conditions of measurement. B is also referred to as the (magnetic) flux density, ϕ/A, where ϕ is the magnetic flux in Webers (Wb) and A is the area normal to the flux.

6.2 Types of Magnetic Order

The chief types of magnetic ordering in solids are diamagnetism, paramagnetism, ferromagnetism, ferrimagnetism and antiferromagnetism. Diamagnetism is a small effect caused by the reaction of the orbiting electrons to an applied magnetic field in accordance with Lenz's law, so that the magnetization and hence the susceptibility are both negative. The magnetic induction is less in the material than it would be in a vacuum with the same field. Typical susceptibilities are about 10^{-5} (for example water, -9×10^{-6}, graphite -8×10^{-5}). Most everyday materials are diamagnetic, but the phenomenon has few uses.

Paramagnetism is a relatively weak effect akin to diamagnetism, but the susceptibility is small and positive: application of a magnetic field to a paramagnet increases the induction beyond what it would be in a vacuum. Paramagnetism is strongly dependent on the temperature, being reduced as the temperature rises. Ferromagnetic materials become paramagnetic above their Curie temperatures (T_c). Little use is made of paramagnetism.

Ferromagnetism, ferrimagnetism and antiferromagnetism are all properties of materials possessing magnetic order, even in the absence of an applied field. If we think of each atom acting like a tiny bar magnet, then in a ferromagnet they are all aligned parallel to each other as in figure 6.1(a). In an antiferromagnet alternate atoms align their magnetic moments antiparallel, so that the net magnetization is zero, as in figure 6.1(b). In a ferrimagnet the magnetic moments on some atoms is less than on others, so that when the two moments align antiparallel, as in an antiferromagnet, there is incomplete cancellation and a net magnetization results as in figure 6.1(c). A ferrimagnet behaves much like a ferromagnet, and the commonest and cheapest magnets are made from ferrimagnetic materials such as the hexaferrite ceramics $BaFe_{12}O_{19}$ (sometimes written $BaO.6Fe_2O_3$) and $SrFe_{12}O_{19}$. Ferrite inductor and transformer cores are also made from ferrimagnets, such as nickel ferrite, $NiFe_2O_4$ (which can be written $NiO.Fe_2O_3$).

The alignment of magnetic moments need not be so simple as in figures 6.1(a), 6.1(b) and 6.1(c), but can be at other, constant angles as in figure 6.1(d), which shows a *canted spin* system. The magnetic order in ferrimagnetic compounds disappears above their Curie points and the magnetic order in antiferromagnets disappears above a characteristic temperature called the Néel temperature (T_N); in both cases the material then becomes paramagnetic. Little use is made of antiferromagnetism.

Nearly all the technically important magnetic materials are ferromagnetic or ferrimagnetic, which is much the same thing in practice, and most people mean this when they say a material is 'magnetic'.

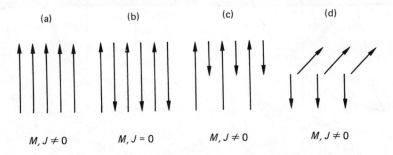

(a)	(b)	(c)	(d)

$M, J \neq 0$ $M, J = 0$ $M, J \neq 0$ $M, J \neq 0$

Figure 6.1   *Alignment of atomic moments in (a) a ferromagnet, (b) an antiferromagnet, (c) a ferrimagnet and (d) a canted spin ferrimagnet*

6.3 The Hysteresis Loop

The most important technological property of a ferromagnet (or ferrimagnet) is hysteresis. Hysteresis comes from the Greek (*hysteros* – later)
and refers to the retardation in the response of a material to a change in
applied field. In figure 6.2(a) is plotted a typical *B–H* diagram for a
ferromagnet as the applied field is increased to a positive value large
enough to saturate the polarization of the material (by equation (6.2), *B*
cannot saturate), then reduced to a negative value large enough to produce
saturation in the reverse direction, then increased back to zero once more.
A symmetrical closed loop known as a hysteresis loop is formed. The
magnetic induction remaining when the applied field is reduced from
saturation to zero is called the *remanence*, B_r. The size of the (negative)
field required to reduce the induction to zero is known as the *coercivity* or
coercive force, H_c. Remanence and coercivity are the properties we shall
try to explain above all as they are of the highest practical significance. If
the applied field is insufficient to saturate the material, then the loop
plotted has a smaller area than the hysteresis loop, and is called an inner
loop. Transformers and inductors should always work below saturation
around inner loops. In this case the working remanent induction is less
than B_r and the coercivity is less than H_c.

From the *B–H* loop we can construct a *J–H* loop by using equation (6.2),
and this has the shape of figure 6.2(b). The magnetic polarization

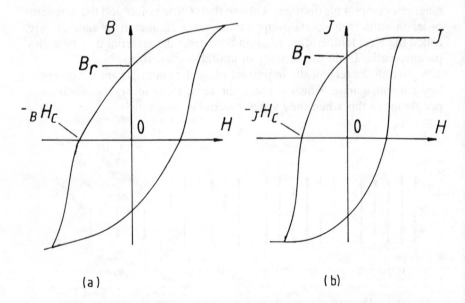

(a)   (b)

Figure 6.2   *(a) A B–H hysteresis loop; (b) a J–H or intrinsic hysteresis loop*

remaining when the field is reduced from saturation to zero is still the remanence, B_r. The field required to reduce the polarization (or magnetization) to zero is called the *intrinsic coercivity* or the *polarization* (or *magnetization*) *coercive force* ($_jH_c$ or $_mH_c$), and cannot be greater than H_c. The polarization, unlike the magnetic induction, saturates in sufficiently large applied fields to a value known as the *saturation polarization*, J_s. The *J–H* loop, often termed the *intrinsic* loop, is more convenient for a theoretical study of magnetization than the *B–H* loop.

Hysteresis loops may be plotted with an apparatus such as that shown in figure 6.3, which is an electronic loop plotter. The sample has a coil closely wound round its centre and is placed between the poles of an iron-cored magnet capable of providing a uniform field within the pole-pieces. If the sample is in contact with the pole-pieces then the sample coil will be enclosing a flux of *BA*, where *A* is the area of the coil (= cross-sectional area of sample). An identical coil to the sample coil, known as the field coil, is placed normal to the field of the magnet. When the flux through the coils changes, an e.m.f. is induced in the coils, which is applied to an electronic integrator, whose output is

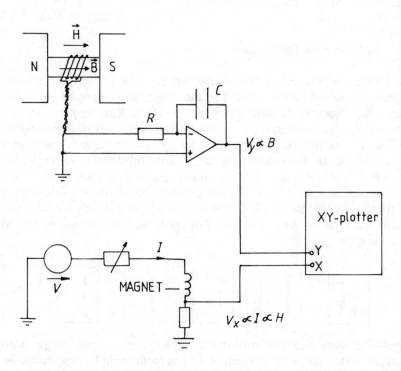

Figure 6.3 *A B–H loop plotter*

$$V = (RC)^{-1} \int N(d\phi/dt)dt$$

$$= N\phi/RC = (NA/RC)B$$

since the flux $\phi = BA$. For the field coil, B is just $\mu_0 H$, while for the sample coil it is $\mu_0 H + J$. If these coils are connected series-opposed, the combined output will be proportional to J. With careful choice of components, the electronic integrator is extremely accurate ($\pm$ 0.1 per cent) and quite sensitive (see problem 6.1).

The work done on the sample by the magnetic field is given by the area of the loop:

$$W_H = \oint H dB = \oint H dJ$$

This work appears in the sample as heat and is the hysteresis loss/cycle. If we would minimize these losses (as in transformer cores), then the loop should have small area, and since a fairly large value of J_s is desirable in a transformer or inductor, this implies that the coercivity be as small as possible. Magnetic materials of low coercivity are called *soft* materials, conversely those of high coercivity are called *hard* materials. (Permanent) magnets are all hard magnetic materials.

6.4   The Saturation Polarization

The saturation polarization is an important intrinsic property of a ferromagnet, because it can be related to the magnetic moment of individual atoms, and because it determines the working flux supplied by the material. Atomic moments are usually measured in units of *Bohr magnetons*, μ_B ($\equiv e\hbar/2m$), corresponding to the orbital magnetic moment of a single electron. In the transition elements of the third row – Sc, Ti, V, Cr, Mn, Fe, Co, Ni, Cu and Zn – the outer electronic structure is $3d^n4s^2$, where n goes from 1 (at Sc) to 10 (at Zn), when the $3d$ orbital is full. The $3d$ electrons obey the rule of *maximum multiplicity*, that is they do not begin to pair until the orbital is half full. For instance, Fe is $3d^6 4s^2$, so the $3d$ electron spins go

The $4s$ electrons make no contribution and there are four unpaired $3d$ electrons, so the magnetic moment of an iron atom might be expected to be $4\mu_B$. Now iron has $J_s = 2.16$ T and from the Periodic Table we find it is bcc

with $a = 2.87$ Å, so that the volume of the unit cell is 2.36×10^{-29} m^3, and its magnetic moment is $J_s a^3$, or 5.1×10^{-29} Tm^3. The Bohr magneton works out to be 9.27×10^{-24} Am^2, or 1.165×10^{-29} Tm^3, so there are $4.4\mu_B$ per unit cell, or $2.2\mu_B$ per atom (bcc means 2 atoms per unit cell), not $4\mu_B$/atom. This is rather surprising, especially as measurements on single iron atoms confirm that the moment is $4\mu_B$. However, in the solid the $3d$ electrons are split into two bands – spin up and spin down – which contain on average 4.8 and 2.6 electrons, for a net moment of $2.2\mu_B$, and there is an average of 0.6 electrons in the $4s$ orbital (paired and so not contributing to the magnetic moment), thus:

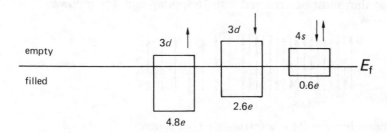

The only other ferromagnetic elements at room temperature are cobalt and nickel, having atomic moments of $1.7\mu_B$ and $0.6\mu_B$ respectively in the solid (compared with $3\mu_B$ and $2\mu_B$ on the isolated atoms). In the other third-row transition elements the magnetic ordering at room temperature is antiferromagnetic (Cr, Mn), diamagnetic (Cu) or paramagnetic (Sc, V, Ti). By alloying iron with cobalt it is possible to achieve a J_s of over 2.5 T, but the improvement is hardly worth the getting. Iron therefore has the highest saturation magnetization of the three room-temperature, elemental ferromagnets, and is also the cheapest; it comes then as no surprise that all large pieces of magnetic material are made mostly of iron. Millions of tons are used for the magnetic circuits of all kinds of generators, motors and transformers: iron is by far the most important magnetic material by weight, though in monetary terms the most important magnetic materials are those used for recording purposes (hard and floppy discs, video and audio tapes).

Looking at ferrimagnets such as $SrFe_{12}O_{19}$, we find a different situation to that in the metallic magnets, because the magnetic atoms (all Fe^{3+} ions) are not in contact with each other, but are separated by oxygen ions, and their electrons do not form energy bands. For this reason the magnetic moments associated with the magnetic atoms are predictable. Fe^{3+} has the outer electronic structure $3d^5$ – there are five electrons in the $3d$ orbitals – which must be unpaired (that is, all spin up) by the law of maximum multiplicity, thus the magnetic moment of an Fe^{3+} ion must be $5\mu_B$:

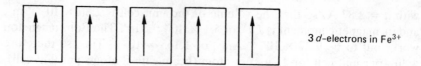

3 d-electrons in Fe^{3+}

The unit cell of strontium hexaferrite is hexagonal, contains two formula units and has $a = 5.87$ Å and $c = 23.0$ Å. The volume of the unit cell is $\frac{1}{2}\sqrt{3}a^2c$, or 6.86×10^{-28} m³ and J_0 is 0.677 T (J_0 is J_s at 0 K), so the magnetic moment of the unit cell is 4.64×10^{-28} Tm³ or 40 μ_B – eight times the moment of an Fe^{3+} ion. Yet there are 24 Fe^{3+} ions in the unit cell, so they must be arranged with 16 spin up and 8 spin down:

Fe^{3+} moments

Strontium hexaferrite is a ferrimagnet, of course.

Most of the Fe^{3+} ions are not making a net contribution to the saturation polarization in the strontium hexaferrite, but efforts to improve matters by substituting other cations into the structure have not succeeded, because the magnetic order breaks down, or the Curie temperature is lowered too much, or the wrong set of iron ions is replaced – see problem 6.4. Attempts to increase J_s have usually foundered in this fashion, though some improvement is often possible.

6.4.1 The Change in Saturation Polarization with Temperature

Ewing attempted in 1900 to explain the behaviour of ferromagnets in terms of tiny magnetic dipoles in the material, which lined up under the influence of an applied field. Such an assembly in reality would behave as a paramagnet, with J_s being strongly dependent on temperature, whereas a ferromagnet like iron, cobalt or nickel has a saturation polarization almost independent of temperature from 300 K down to 0 K (for example, in cobalt $J_s = 1.76$ T at 300 K and $J_0 = 1.82$ T, only 3 per cent more).

In 1907, Weiss introduced two new concepts into Ewing's theory and Langevin's then new theory of paramagnetism. These concepts were the molecular field and the magnetic domain; it is doubtful if any ideas have contributed more to the study of magnetism than these. Domains will be discussed in some detail later and Langevin's theory of paramagnetism will be more fully described in the study of dielectrics, since it applies as well to electric dipoles as magnetic. Essentially it says that if a field is applied to a medium containing N dipoles/m^3, each with a moment, η, they will tend to align with the field, though disturbed by random thermal energy. The resulting polarization is

$$J = N\eta L(x)$$

$L(x)$ is the Langevin function, $\coth(x) + 1/x$, $x \equiv \eta H/k_B T$ and η is in Tm^3.

Weiss suggested that there was in a ferromagnet an internal, or molecular, field acting on the atomic moments to produce alignment even without an applied field. This molecular field, H_w, is proportional to the polarization:

$$\mu_0 H_w = \lambda J$$

where λ is the molecular field, or Weiss, constant. H_w is very large indeed: take an iron atom with a magnetic moment of $2.2\mu_B$, whose magnetic energy in the molecular field is then $2.2 \times 1.165 \times 10^{-29} H_w$, which is approximately equal to $k_B T_c$, since $T_c = 1043$ K, we find $H_w \approx 560$ MA/m ($\mu_0 H_w \approx 700$ T), a far larger field than can be obtained in a laboratory. (A typical superconducting solenoid will give 10 T and a very expensive, specialized, research installation maybe 25 T.)

Weiss's idea was to put $x \equiv \eta H_w/k_B T$ into the Langevin function to give the polarization as a function of time. Though the fit was better, it was still not very good. After the quantum theory was discovered, Brillouin derived an equation which was an almost exact fit for iron, cobalt and nickel:

$$J = N\eta\tanh(\eta\lambda J/k_B T)$$

or

$$J/N\eta \equiv j = \tanh(\eta^2 \lambda N j/k_B T)$$

which can be written

$$j = \tanh(j/t) \tag{6.4}$$

where t is the reduced temperature, T/T_c ($T_c \equiv \eta^2 N\lambda/k_B$), and j is the reduced polarization, J/J_0 ($J_0 \equiv N\eta$). When $t = 1$, $j = 0$, so the polarization vanishes at T_c, as $t = 1$ corresponds to $T = T_c$. A plot of (6.4) is shown in figure 6.4. When the reduced temperature is less than about 0.4, the saturation polarization is close to its value at 0 K. At room temperature $t \approx 0.3$ for iron, so $J_s = J_0$.

Ferrimagnets tend not to follow equation (6.4) very closely, and often the change in polarization with temperature is nearly linear from 0 K to the Curie temperature. The reason for this is that, although the opposing sublattices individually obey (6.4) or nearly so, the resultant of the two does not and is almost linear except near $T = 0$ and $T = T_c$, as in figure 6.5. J_s tends to be low in ferrimagnets because of this as well as the opposed moments.

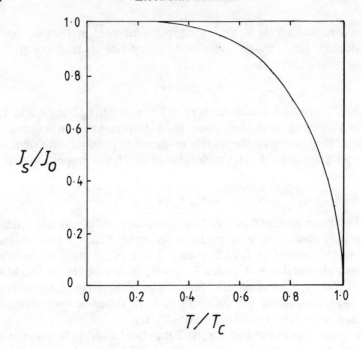

Figure 6.4 J_s/J_0 against T/T_c; the tanh law

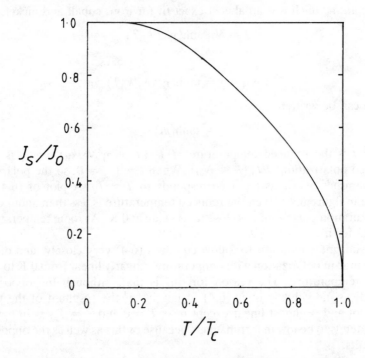

Figure 6.5 *j against t for a ferrimagnet*

6.5 Anisotropy Energy

Any magnetic material will be easier to magnetize in some directions rather than others, which is the origin of magnetic anisotropy energy. The three main sources of this anisotropy are:

(1) Crystal structure, giving rise to *magnetocrystalline anisotropy*
(2) Specimen shape, giving rise to *shape anisotropy*
(3) *Strain anisotropy*.

6.5.1 Magnetocrystalline Anisotropy

When a single crystal of a ferromagnet (or ferrimagnet) is magnetized by an applied field it is found that the crystal is more readily magnetized in some directions than others, as shown for cobalt in figure 6.6. The energy per unit volume required for magnetizing the specimen is just $\int H\mathrm{d}J$, the area between the polarization curve and the polarization axis. So in the case of cobalt, the magnetizing energy is rather small in the <00.1> directions (the easy directions), but about $\frac{1}{2}J_s H_s$ in the <10.0> directions (the hard directions), where J_s, the saturation polarization, is 1.8 T and H_s is the saturating field in the <10.0> directions, about 0.65 MA/m. The difference in magnetizing energies, which is about 550 kJ/m³, is known as the magnetocrystalline anisotropy energy (per m³). It may be expressed for uniaxial crystals, such as the hexagonal materials cobalt and barium hexaferrite, in the form

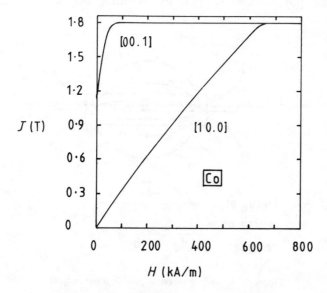

Figure 6.6 *Magnetic anisotrophy in single-crystal cobalt*

$$U_A = \sum_1^n K_n \sin^{2n}\vartheta \qquad (6.5)$$

where K_n is the nth magnetocrystalline anisotropy constant, and ϑ is the angle between the magnetization vector and the c-axis (the [00.1] direction). Seldom are more than two terms of equation (6.5) needed, and often one will do. If the magnetization in cobalt is along <10.0>, then $\vartheta = 90°$, and $U_A = K_1 + K_2 = 550$ kJ/m^3. Measurements in other crystallographic directions are necessary to find K_1 and K_2 separately: experimentally, $K_1 \approx 410$ kJ/m^3 and $K_2 \approx 100$ kJ/m^3. We shall use only K_1, as it usually gives sufficient accuracy and the magnetocrystalline anisotropy energy is then just

$$U_A = K_1 \sin^2\vartheta \qquad (6.6)$$

Cubic crystals require a more complicated expression which we shall not need as their magnetocrystalline anisotropy energies are relatively small (for example, K_1 for iron is only a tenth of cobalt's).

6.5.2 Shape Anisotropy

To explain the origin of shape anisotropy needs the introduction of the concept of the *demagnetizing field*. Consider the bar magnet in figure 6.7, which is magnetized to saturation along its long axis, producing poles at either end as shown. There are field lines running from north pole to south pole not only outside but also inside the magnet. These internal field lines oppose the polarization, so that the resulting magnetic induction is reduced:

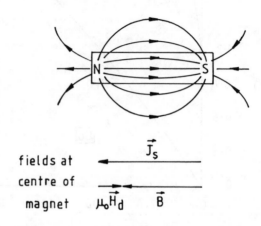

Figure 6.7 *Lines of magnetic field in a bar magnet*

$$B = J_s - \mu_0 H_d$$

where H_d is the strength of the field in the magnet due to its own poles, called the *demagnetizing field*, for apparent reasons. It is reasonable to suppose that when the poles are very far apart, as in a long, needle-like magnet, the demagnetizing field is very small, as in figure 6.8(a). By the same token, if the specimen is very short and fat, so that the poles are close together, then the demagnetizing field is at a maximum, whose magnitude is J_s/μ_0, and the specimen has a resultant induction of zero, as in figure 6.8(b).

In a specimen of intermediate dimensions the demagnetizing field will be somewhere between zero and $J_s\mu_0$, for example a sphere has a demagnetizing field of $J_s/3\mu_0$, and this is obviously independent of the orientation. In general, the demagnetizing field is given by

$$H_d = - N_d J_s$$

where N_d is the *demagnetizing factor*, a tensor quantity in general.

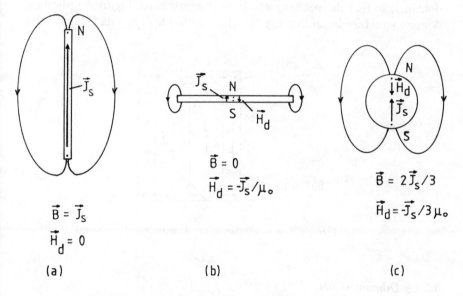

(a)   (b)   (c)

Figure 6.8   *(a) B, H_d and J_s for a needle-shaped particle; (b) B, H_d and J_s for a plate-like particle; (c) B, H_d and J_s for a spherical particle*

As a consequence of the demagnetizing field a magnet has a *magnetostatic energy* per unit volume, U_M, given by

$$U_M = - \tfrac{1}{2} H_d J_s = \tfrac{1}{2} N_d J_s^2 \tag{6.7}$$

Thus, if we magnetize a thin disc normal to its surface, it will have a magnetostatic energy of $J_s^2/2\mu_0$, whereas if we were to magnetize it in the plane of the disc the magnetostatic energy would be zero. Because a disc possesses shape anisotropy, the magnetization has a preferred direction – the plane of the disc. Spheres alone have no shape anisotropy, so we should expect the magnetization in practice to have easy and hard directions, regardless of other anisotropies. Here we shall leave our study of energy anisotropy and deal with strain anisotropy later.

6.6 Magnetic Domains

The second of Weiss's revolutionary ideas and the most important was that of the magnetic domain. It is common experience that a piece of iron is not normally very much magnetized, though it can easily be made so with a solenoid or magnet. The reason for there being no observable magnetization is that the material contains domains in each of which the magnetization is directed uniformly, but is oppositely directed in neighbouring domains, so that the net magnetization is nearly zero. Figure 6.9 shows a domain structure in an iron crystal which would lead to this result.

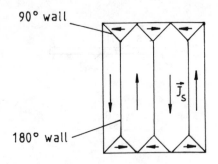

Figure 6.9 *Closure domains in an iron crystal*

6.6.1 Domain Walls

Magnetic domains are separated by *domain walls* (often called *Bloch walls*), which are regions of high energy density, since neighbouring atomic moments are not aligned parallel, as may be seen in figure 6.10. Neighbouring atomic moments are aligned parallel by *exchange energy*, given by

$$U_E = -2J_{ex}S_i.S_j \tag{6.8}$$

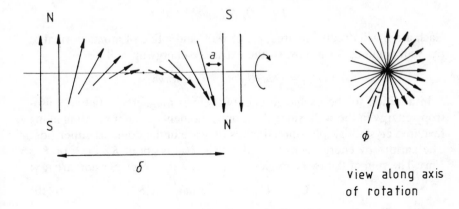

view along axis
of rotation

Figure 6.10 *The rotation of moments in a 180° domain wall*

where S_i, S_j are the spins (in effective numbers of Bohr magnetons) of atoms i, j, and J_{ex} is the exchange integral, an energy term which is about 0.01 eV for bcc iron, and about $-$ 0.001 eV in fcc iron (which is antiferromagnetic, hence the minus sign). The exchange energy is a quantum effect of Coulombic origin. (6.8) is sometimes called the Heisenberg Hamiltonian. J_{ex} is given by

$$J_{ex} = K_B T_c / 2zS(S + 1) \qquad (6.9)$$

where T_c is the Curie temperature and z is the number of nearest neighbours of an atom in the solid.

Consider the 180° domain wall (in uniaxial materials, only 180° walls are possible, but in cubic materials, such as iron, 90° walls are found too) shown in figure 6.10. By (6.8), the exchange energy for a pair of atoms in the wall is

$$U_E = - 2J_{ex}S^2 \cos\phi$$

where S is the spin on an atom and ϕ is the angle between the moments of neighbouring atoms. Now ϕ is small as the wall thickness is large compared with the atomic spacing, so $\cos\phi \approx 1 - \frac{1}{2}\phi^2$ and

$$U_E = - 2J_{ex}S^2(1 - \tfrac{1}{2}\phi^2)$$

But if the spins had been parallel, U_E would have been $- 2J_{ex}S^2$, so the increase in energy for an atom in the wall due to spin misalignment is $J_{ex}S^2\phi^2$. Now the wall thickness is δ and the distance between neighbouring atoms is a, so the wall is δ/a atoms thick and the magnetization rotates through 180° over this number of atoms. Thus $\phi = \pi a/\delta$ radians and the exchange energy/atom is

$$U_E = J_{ex}S^2(\pi a/\delta)^2$$

Each atom will occupy an area, a^2, of wall, and a line of atoms normal to the wall contains δ/a atoms, so the exchange energy/m^2 will be

$$U_{ex} = (\delta/a)U_E/a^2 = J_{ex}S^2\pi^2/a\delta$$

In addition to the exchange energy there is magnetocrystalline aniso-tropy energy in the wall, since the atomic moments are not pointing along favoured crystallographic directions as they are in the domains either side. The anisotropy energy density of the wall, U_A, is about K_1 J/m^3, or $K_1\delta$ J/m^2. The sum of the two energies will be the wall energy per unit area, γ:

$$\gamma = U_{ex} + U_A = J_{ex}S^2\pi^2/a\delta + K_1\delta \qquad (6.10)$$

A competition is going on between the exchange energy, which tries to increase the wall thickness, and the anisotropy energy, which tries to keep the wall as thin as possible; a compromise is achieved when the two energy contributions are equal, as may be seen by differentiating (6.10) and equating to zero:

$$d\gamma/d\delta = -J_{ex}S^2\pi^2/a\delta^2 + K_1$$

which is zero when

$$\delta = \sqrt{(J_{ex}S^2\pi^2/K_1a)}$$

Substituting for δ in (6.10) leads to

$$\gamma = 2\sqrt{(J_{ex}K_1S^2\pi^2/a)} = 2K_1\delta$$

J_{ex} can be replaced using (6.9) to give

$$\gamma = 2\sqrt{(k_BT_cK_1S^2\pi^2/2zS[S+1]a)}$$

$$\approx \sqrt{(2k_BT_cK_1/a)} \qquad (6.11)$$

and

$$\delta = \gamma/2K_1 \approx \sqrt{(k_BT_c/2K_1a)} \qquad (6.12)$$

The approximations result because $zS(S+1) \approx \pi^2S^2$. Putting into (6.11) the values for iron ($K_1 = 40$ kJ/m^3, $T_c = 1043$ K, $a = 2$ Å) leads to $\gamma = 2.4$ mJ/m^2 and $\delta = 30$ nm – about the same as the experimental figures. As we shall see, the wall energy is important for understanding the size of the hysteresis losses and the coercivity.

6.6.2 *Single-domain Particles*

Louis Néel introduced this concept in about 1943, though Stoner and Wohlfarth presented the idea in 1947, seemingly unaware of Néel's work. It provided a powerful stimulus for research into permanent magnets for two decades thereafter. The theory suggests there is a critical size for a magnetic particle below which domain walls will not form, so that the particle contains but a single domain.

Consider the sphere shown in figure 6.11(a), which contains no domain wall and has a magnetostatic energy/m³ given by

$$U_{\mathrm{m}} = \tfrac{1}{2}N_{\mathrm{d}}J_s^2 = J_s^2/6\mu_0$$

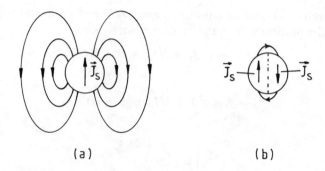

(a)   (b)

Figure 6.11   *(a) The external field of a spherical single-domain particle; (b) the external field after the formation of a domain wall*

as $N_{\mathrm{d}} = 1/3\mu_0$. If a domain wall now forms in the middle, as in figure 6.11(b), then the magnetostatic energy is reduced (the demagnetizing factor of half a sphere is less than that of a sphere) – roughly by half, so that $U_{\mathrm{M}} \approx 0.08J_s^2/\mu_0$. If the radius of the sphere is r, then the reduction in magnetostatic energy by the domain wall is $0.08J_s^2/\mu_0 \times \tfrac{4}{3}\pi r^3$, while the area of the wall is πr^2 and the particle's wall energy is $\pi r^2\gamma$. Thus wall-formation is energetically favorable if

$$\pi r^2\gamma > 0.08J_s/\mu_0 \times \tfrac{4}{3}\pi r^3$$

or

$$r > 9\mu_0\gamma/J_s^2$$

In iron, where $\gamma = 2.4$ mJ/m² and $J_s = 2.16$ T, the critical radius works out at 6 nm, compared with a wall thickness of 30 nm: so we would not expect to find spherical, single-domain particles of iron. However, in strontium hexaferrite γ is 9 mJ/m² – larger than in iron – while J_s is much less at 0.48 T, so that the critical radius is about $\tfrac{1}{2}$ μm. Commercial ceramic magnets are thought to comprise single-domain particles, which makes their coercivity higher than when they contain domain walls.

6.6.3 *Hysteresis of Single-domain Particles*

Let us look at a spherical, uniaxial, single-domain particle which is hexagonal with the *c*-axis being the easy axis, as in figure 6.12. Suppose the applied field is directed at an angle, ϕ, to the *c*-axis as shown so that the polarization rotates in the direction of the field by an angle, ϑ. The magnetic energy per unit volume is made up of two terms:

$$U = U_A + U_H$$

U_A is the magnetocrystalline anisotropy energy (given by 6.6) and U_H is the energy of the particle due to the applied field:

$$U_H = - \mathbf{H}.\mathbf{J}_s = HJ_s\cos(\phi + \vartheta)$$

Thus

$$U = K_1\sin^2\vartheta + HJ_s\cos(\phi + \vartheta) \tag{6.13}$$

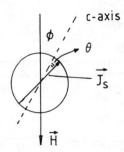

Figure 6.12 *A uniaxial, single-domain particle in an applied field*

To simplify matters, consider the limiting case where $\phi = 0°$, that is, the polarization starts off at 180° to the applied field. Then (6.13) becomes

$$U = K_1\sin^2\vartheta + HJ_s\cos\vartheta \tag{6.14}$$

Dividing by $2K_1$

$$U/2K_1 \equiv u = \tfrac{1}{2}\sin^2\vartheta + h\cos\vartheta \tag{6.15}$$

where $h \equiv H/H_A$ and H_A ($\equiv 2K_1/J_s$) is the *anisotropy field*. Differentiating (6.15) with respect to ϑ gives

$$\partial u/\partial\vartheta = u' = \sin\vartheta\cos\vartheta - h\sin\vartheta \tag{6.16}$$

Setting $u' = 0$ will give turning points for u and hence U, which are when $\sin\vartheta = 0$ ($\vartheta = 0°$ or $\vartheta = 180°$) and when $h = \cos\vartheta$. To find whether these are maxima or minima we can differentiate (6.16) with respect to ϑ, getting

$$u'' = \cos^2\vartheta - \sin^2\vartheta - h\cos\vartheta$$

When $\vartheta = 0°$, $u'' = 1 - h$, which is positive if $h < 1$, that is, u is a minimum. When $\vartheta = 180°$, $u'' = 1 + h$, positive if $h > -1$, another minimum. $h = \cos\vartheta$ gives $u'' = h^2 - 1$, negative if $|h| < 1$, that is, a maximum. Thus the stable values of ϑ are 0° and 180°, so that the polarization remains at $\vartheta = 0°$ until $h = 1$ (that is, $H_A = H$), when it rotates through 180° and aligns with the field. Further increase of field does not cause any change in polarization of course. The hysteresis loop of the single-domain particle thus looks like figure 6.13(a), and is termed a *square*

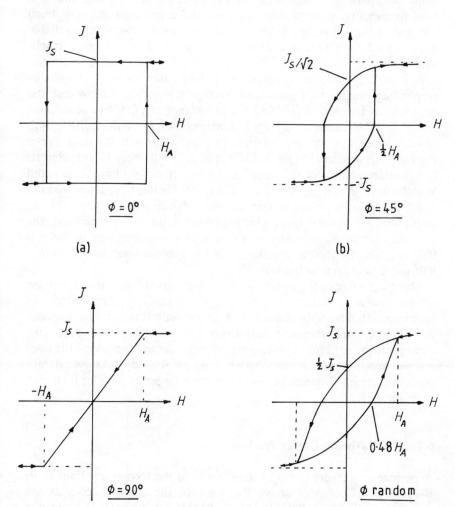

Figure 6.13 *Hysteresis loops for the particle in figure 6.12: (a) with $\phi = 0°$, (b) with $\phi = 45°$, (c) with $\phi = 90°$ and (d) with ϕ random*

loop. The coercivity is here equal to H_A, the anisotropy field. A square-loop material has a remanence equal to the saturation polarization, the highest possible value; and not only is B_r as high as it can be, but so also is the coercivity. The *squareness factor* of a hysteresis loop is the ratio J_s/B_r, so a square loop is one with a squareness factor of one.

Magnetocrystalline anisotropy fields can be very large, so that the coercivity of an ideally oriented, single-domain particle should theoretically be equally large. For instance, $SrFe_{12}O_{19}$ has $K_1 = 350$ kJ/m^3 and $J_s = 0.48$ T, making $H_A = 1.5$ MA/m (≈ 19 kOe). $SmCo_5$ has an anisotropy field perhaps as high as 20 MA/m (≈ 250 kOe) – so high that it is hard to measure. In both cases the observed coercivities are very much lower than the anisotropy field; however, we have established a reliable upper limit for the intrinsic coercivities of magnets due to magnetocrystalline anisotropy.

If $\phi = 45°$, then the hysteresis loop looks like figure 6.13(b), corresponding roughly to a particle of average orientation. In this case the coercivity is lowered to $\frac{1}{2}H_A$ and the remanence to $H_s/\sqrt{2}$ (a squareness factor of 0.707), compared with a remanence of J_s for the particle with $\phi = 0°$. When $\phi = 45°$, the polarization never aligns with the field. When the easy axis is normal to the field, that is, $\phi = 90°$, there is no hysteresis (see problem 6.5) and the 'loop' looks like figure 6.13(c). Stoner and Wohlfarth (*Phil. Trans. Roy. Soc.*, **240A**, 599 (1948)) have calculated the hysteresis loop for an assemblage of randomly-oriented particles with the result shown in figure 6.13(d). One important result of this work is that the theoretical intrinsic coercivity of a random array of magnetic particles is $0.48 H_A$. The theoretical remanence of the random array is exactly $\frac{1}{2}J_s$ – half the maximum possible value.

The great value of the Stoner–Wohlfarth theory is that it is *independent of the source of the anisotropy*: it applies equally to shape, strain or magnetocrystalline anisotropies. For shape anisotropy, if we consider disc-like particles, then the anisotropy field's magnitude is J_s/μ_0 (the demagnetizing field for polarization perpendicular to the plane of the disc) and the greatest possible coercivity is J_s/μ_0 for aligned, and $0.48 J_s/\mu_0$ for random particles. Once more, these are much larger coercivities than seen in practice.

6.7 The Maximum Energy Product

A permanent magnet operates somewhere in the second quadrant of its *B–H* loop, usually just above the knee of the curve, known as the *demagnetization curve*. The quantity $- BH$ (H is negative) is an indication of (but does not equal) the energy stored in the field per unit volume of magnet. The demagnetization curve for an alnico magnet is shown in figure

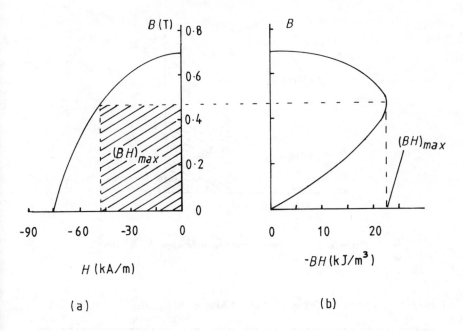

Figure 6.14   *(a) The demagnetization curve for an alnico magnet; (b) − BH against H from (a)*

6.14(a) and the plot of $-BH$ against H derived from this loop in figure 6.14(b). The $-BH$ against H curve has a maximum at a particular value of H, known as $(BH)_{max}$, which is a figure of merit for a permanent magnet material: the higher $(BH)_{max}$, the smaller is the volume of the magnet required to produce a given flux.

How can $(BH)_{max}$ be increased? There are just two ways – increase the coercivity and the remanence – but they must both be increased at the same time, for a material with a very high remanence and a low coercivity will not have high $(BH)_{max}$. However, there is a limit to $(BH)_{max}$. Consider the demagnetization curve of a material with a square intrinsic hysteresis loop and infinite intrinsic coercivity, as in figure 6.15. The highest achievable B_r is J_s, and if $_jH_c$ is infinite, then $_BH_c$, which is just the coercivity, must be B_r/μ_0 $(= J_s/\mu_0)$. The demagnetization curve in this case is a straight line and the $(BH)_{max}$ point is halfway down where $B = \frac{1}{2}B_r$ and $H = -\frac{1}{2}B_r/\mu_0$, so $(BH)_{max}$ is $\frac{1}{2}B_r^2/\mu_0$, which is at the utmost, $\frac{1}{4}J_s^2/\mu_0$. No matter what the magnet, it can never exceed this value. In table 6.1, $(BH)_{max}$ for diverse magnet materials is compared to the theoretical maximum; in some cases, when $_jH_c$ is large, it is close.

The properties of magnets may be improved by orienting the magnetic particles so that their preferred axes of magnetization are aligned. For

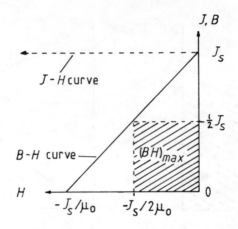

Figure 6.15 *The demagnetization curve when $_jH_C$ is infinite*

Table 6.1 Properties of Some Permanent Magnet Materials

Magnet	Composition[*]	B_r (T)	H_c[†] (kA/m)	$(BH)_{max}$ (kJ/m³)	$\frac{1}{4}B^2_r/\mu_0$ (kJ/m³)
Alnico 2[a]	Al9Ni20Co18Cu3Ti6	0.7	72	16	97
Alnico 9[b]	Al8Ni15Co29Cu3Ti5	1.1	110	75	241
Feroba 1[c]	$MO.6Fe_2O_3$	0.20	135	7.5	8.0
Feroba II[d]	$MO.6Fe_2O_3$	0.40	260	30	32
Feroba[e]	$MO.6Fe_2O_3$ + resin	0.17	110	5	5.7
Feroba[f]	$MO.6Fe_2O_3$ + resin	0.25	170	10	12
RE–Cobalt	$SmCo_5$	0.9	680	160	161
RE–Cobalt	Sm_2Co_{17}	1.1	560	240	241
Fe–Cr–Co[i]	Fe57Cr22Co18Mo3	1.58	72	90	497
ESD 50[g]	Fe100	0.88	58	26	175
Steel[h,j]	Fe93W6Cr1	1.05	5.2	2.4	220
Cunife[i,j]	Cu60Ni20Fe20	0.57	35	11	65
Platinax[j]	Pt77Co23	0.64	400	76	81

[a] Isotropic. Balance of composition is iron.
[b] Anisotropic. Balance of composition is iron.
[c] Isotropic. M is Ba or Sr.
[d] Anisotropic. M is Ba or Sr.
[e] Isotropic. Resin bonded, hence lower B_r.
[f] Anisotropic. Resin bonded.
[g] Elongated Single Domain iron magnets. Made by 𝒢𝒠™ of USA.
[h] Original horseshoe magnet material.
[i] Somewhat ductile.
[j] No longer manufactured.
[*] In wt% unless given as a chemical formula, such as $SmCo_5$.
[†] This means $_BH_c$, *not* $_JH_c$

alnico magnets, which rely on the shape anisotropy of a ferromagnetic phase in a paramagnetic matrix, this means having the long axis of the ferromagnetic particles in the direction of the magnet's polarization, which can be achieved by cooling them in a magnetic field after casting.

Ceramic magnets are made from fine powders which can be oriented by vibrating them in an alternating magnetic field before pressing to shape in a die. In this way oriented hexaferrite magnets achieve a doubling of remanence compared with the unoriented magnets and hence a quadrupling of maximum energy product.

6.8 Hysteresis in Multi-domain Magnetic Materials

In practice, magnets are never as good as theory says they should be, and one reason for this is that they contain domain walls. Once a domain wall is nucleated it can move through the material under the influence of an applied field and reverse the magnetization as it passes; unless something impedes this motion, the coercivity is nil. What can impede the motion of domain walls is a non-magnetic inclusion, such as a precipitate or a void. Non-magnetic inclusions pin domain walls by reducing the total wall energy in the magnet. Consider the spherical void shown in figure 6.16(a), which is intersected by a 180° domain wall. If a magnetic field is applied in

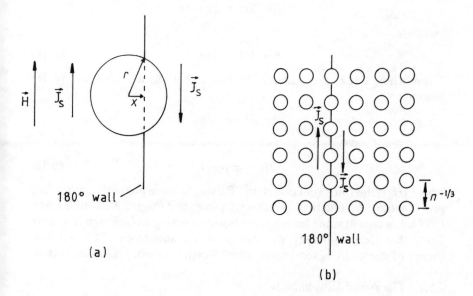

Figure 6.16 (a) A spherical void intersected by a 180° wall; (b) a regular array of voids intersected by a 180° wall

the direction shown, the wall will move to the right a distance δx. Now the original area of 'missing' wall is given by

$$A = \pi(r^2 - x^2)$$

Differentiating

$$\delta A = 2\pi x \delta x$$

where δA is the increase in wall area, making a gain in wall energy, δU_w:

$$\delta U_w = -\gamma \delta A = 2\pi x \delta x \tag{6.17}$$

Suppose that there are n identical voids/m^3 as in figure 6.16(b) so that the volume of magnet containing one void is $1/n$ m^3. Imagine a cube of this volume, so that its side length is $n^{-\frac{1}{3}}$, and the area of domain wall in this cube will be $n^{-\frac{2}{3}}$ m^2. Moving this wall under the influence of the applied field by a distance δx will reduce the magnetic energy by $2HJ_s\delta V$, where δV is the volume in the cube undergoing magnetization reversal, given by

$$\delta V = n^{-\frac{2}{3}}\delta x$$

so that the reduction in magnetic energy, δU_m, is

$$\delta U_m = 2HJ_s n^{-\frac{2}{3}}\delta x$$

At equilibrium this energy must equal the increase in energy, δU_w, given by (6.17), so that

$$2HJ_s n^{-\frac{2}{3}}\delta x = 2\pi\gamma x\delta x$$

and thus

$$x = HJ_s n^{-\frac{2}{3}}/\pi\gamma \tag{6.18}$$

But when $x = r$, the wall has broken free from the void and (6.18) becomes

$$r = {}_jH_cJ_s n^{-\frac{2}{3}}/\pi\gamma$$

or

$${}_jH_c = \pi\gamma rn^{\frac{2}{3}}/J_s \tag{6.19}$$

where ${}_jH_c$ is the (intrinsic) coercivity. Putting in some plausible figures (see problem 6.6), we find that the coercivity in a steel magnet is about 4 kA/m (50 Oe), a typical value for the early horseshoe magnets, though it is more likely that their coercivity was due to strain anisotropy. The inclusion theory of coercivity is sometimes called Kersten's theory, after its author.

6.8.1 The Initial Susceptibility

In small fields, domain walls will not move bodily through the material, but will remain in sites of minimum energy, that is, at a pinning site, and will accommodate the increase in energy cause by the field by bulging so as to increase the volume of favourably oriented polarization. A 180° domain wall is seen in figure 6.17, pinned at two sites separated by a distance y and it bulges into a cylindrical surface of length y (y is the average spacing between pinning sites). The depth of the bulge is x, so that the volume of

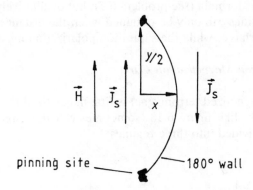

Figure 6.17 *Bulging of a pinned 180° domain wall*

the bulge, δV, is about $xy^2/2$ ($x << y$) and the reduction in magnetostatic energy is $2HJ_s\delta V$. The increase in wall area is

$$\delta A \approx 2y\sqrt{(x^2 + \tfrac{1}{4}y^2)} - y^2 \approx 2x^2$$

So the increase in wall energy, $\gamma\delta A$, is $2\gamma x^2$. Equating the two energies gives

$$2\gamma x^2 = 2HJ_s\delta V = HJ_sxy^2$$

leading to

$$x = \tfrac{1}{2}HJ_sy^2/\gamma$$

and

$$\delta V = \tfrac{1}{2}xy^2 = \tfrac{1}{4}HJ_sy^4/\gamma$$

The change in magnetic moment caused by this bulge will be just $2J_s\delta V$, and if there are n such pinning sites/m³, the total change in polarization (δJ) will be $2nJ_s\delta V$, that is

$$\delta J = 2nJ_s \times \tfrac{1}{4}HJ_sy^4/\gamma$$

$$= \tfrac{1}{2}nHJ_s^2y^4/\gamma$$

Now the number of pinning sites/m³, n, will be proportional to y^{-3}, giving

$$\delta J \approx HJ_s^2y/\gamma$$

so

$$\delta J/\mu_0 H = J_s^2 y/\gamma\mu_0$$

But $\delta J/\mu_0 H = M/H = \chi_i$, so $\chi_i = J_s^2 y/\gamma\mu_0$.

χ_i is the initial susceptibility, and the constant of proportionality has been set to two, for the sake of convenience. We should not expect accurate results from this formula (see problem 6.7), but qualitatively it shows that high permeabilities can only be obtained when the distance between wall pinning sites is large, while the saturation polarization must be high.

6.8.2 The Initial Magnetization Curve

When an unmagnetized ferromagnet is first magnetized to saturation, the J–H curve looks like figure 6.18 (sometimes called a virgin curve). The graph can be divided into three regimes:

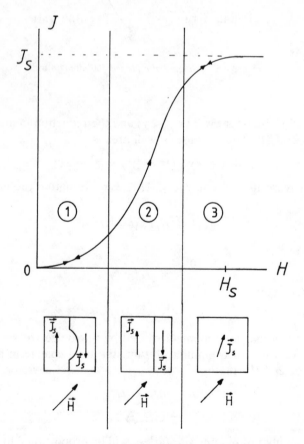

Figure 6.18   *The initial J–H curve (the virgin curve)*

(1) The part near the origin, which has relatively low susceptibility
(2) A region of relatively high susceptibility in which nearly all the polarization is acquired
(3) The last part where there is little change in polarization and low incremental susceptibility (dJ/dH).

In the first part of the curve, the change in polarization is caused by wall bulging, so there is little hysteresis when the field is reversed. In the second part, the walls have been driven from their pinning sites; motion of these walls is easy and dJ/dH is large. Field reversal in this region leads to considerable hysteresis. In the final part of the curve, the walls have been largely eliminated from the crystallites of the material and the polarization increases as a result of rotation against the forces of magnetocrystalline or shape anisotropy. Compared with wall motion, this process is difficult and dJ/dH is small. In this region, the curve is almost reversible.

6.9 Magnetostriction

When a magnetic material is magnetized, it changes its dimensions and this is called magnetostriction. In some materials the magnetostriction is positive and the material expands in the direction of magnetization (and contracts transversely); in others there is contraction in the direction of magnetization (and expansion transversely) and the magnetostriction is negative. In materials with positive magnetostriction, the magnetization is increased by tensile stress and in those with negative magnetostriction it is reduced; in both cases the changes in the hysteresis loop are pronounced, as shown in figure 6.19.

Magnetostriction causes an audible hum in transformer cores, which can be a bother, besides reducing the permeability. In nickel and some materials recently discovered, based on rare earth elements, magnetostriction is used to convert electrical energy to sound energy and vice versa: nickel has been used in sonar and ultrasonic transducers for more than fifty years.

In general the magnetostriction is a function of crystallographic direction, but in polycrystalline materials, such as those used in practice, the effect is isotropic and the energy/m^3 associated with magnetostriction is given by

$$U_s = 1\tfrac{1}{2}\lambda_s T \sin^2\vartheta$$

where λ_s is the magnetostriction coefficient, that is, the relative change in length when the material is saturated, T is the tensile stress and ϑ is the angle between T and J. Thus if a field is applied to a material containing

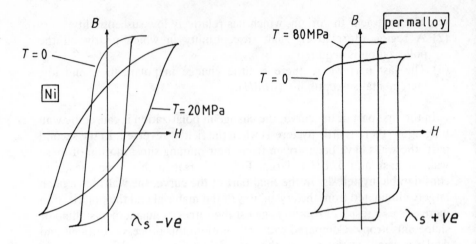

Figure 6.19 *Hysteresis loops for magnetostrictive materials*

stressed regions (for example, around precipitates, or cold-worked samples), the total energy per unit volume is of the form

$$U_T = 1\tfrac{1}{2}\lambda_s T\sin^2\vartheta + HJ_s\cos\vartheta$$

which is similar to equation (6.14), with $1\tfrac{1}{2}\lambda_s T$ replacing K_1. The stress anisotropy field is $3\lambda_s T/J_s$. In iron, λ_s is 4×10^{-6} and J_s is 2.16 T, so that to achieve a coercivity of 5 kA/m would require a stress of 9×10^8 Pa, corresponding to a strain of $\tfrac{1}{2}$ per cent – a reasonable value. Stress anisotropy probably accounts for the coercivity of the obsolete steel magnets.

Problems

6.1. An electronic integrator is used to measure the susceptibility of a paramagnetic sample which is a cylinder 12 mm in diameter. A coil of 1000 turns is closely wound round the centre of the sample which is then placed within the poles of a magnet so that its axis is parallel to the magnet's field. The sample coil is connected in series opposition with a field coil which has exactly the same area–turns product. When the magnet's field is increased from zero to 100 kA/m, the integrator's output is 0.1 mV. If the integrator's time constant (RC) is 1 ms, find the sample's susceptibility.
[*Ans.* 7×10^{-6}]

6.2. Second quadrant J–H data from a permanent magnet material are given in the table below. Plot the J–H and B–H second quadrants from the data given and estimate $_jH_c$, H_c, B_r and $(BH)_{max}$. What do you think the material is?

J (T)	0.24	0.23	0.22	0.21	0.19	0.16	0.13	0.09	$-$ 0.02
$-H$ (MA/m)	0.00	0.04	0.08	0.12	0.16	0.20	0.22	0.23	0.24

[Ans. 0.24 MA/m, 0.16 MA/m, 0.24 T, 9.6 kJ/m³]

6.3. Equation (6.4) is not easy to solve in the form given. Show that it can be written

$$t = 2j/\ln[(1 + j)/(1 - j)]$$

This form is more readily soluble for either variable. Hence find J_s at 300 K for Ni for which $J_0 = 0.641$ T and $T_c = 627$ K.
[The experimental value is 0.610 T.]
[Ans. 0.62 T]

6.4. $SrFe_{12}O_{19}$ is a uniaxial, hexagonal ferrimagnet with a magnetic structure as outlined in section 6.4. An attempt is made to increase J_0 (and hence J_s) by substituting Al^{3+} for Fe^{3+} to form $SrFe_{10}Al_2O_{19}$. Measurements on single crystals at 0 K show that the lattice parameters are unchanged but that J_0 is reduced to 83 per cent of the value for the unsubstituted compound. What is the distribution of the Al^{3+} ions?

Measurements of K_1 are made on both compounds at 300 K and it is found that $K_1 = 0.4$ MJ/m³ in each case. The Curie temperatures are found to be 750 K for $SrFe_{12}O_{19}$ and 600 K for $SrFe_{10}Al_2O_{19}$. Assuming dJ_s/dT is linear from 0 K to T_c, what would you estimate for J_s and H_A for each compound at 300 K? Which compound is likely to have the higher coercivity? Estimate $(BH)_{max}$ at 300 K for polycrystalline samples of each formula having randomly oriented grains.
[Ans. 0.4 T and 0.28 T; 2 MA/m and 2.86 MA/m; 8 kJ/m² and 4 kJ/m³]

6.5. Use equation (6.13) to find the hysteresis loop of a single-domain uniaxial particle when $\phi = 90°$.

6.6. A sample of steel contains 1 per cent by volume of a non-magnetic phase comprising identical, spherical particles of diameter 150 nm. Use Kersten's inclusion theory to estimate the coercivity of the sample.
[Take the wall energy to be 2.4 mJ/m², and J_s to be 1.5 T.]
[Ans. 1200 A/m]

6.7. Domains are pinned in a material by inhomogeneities having an average spacing of 1 μm. If the domain wall energy, γ, is 1 mJ/m² and J_s is 2 T at 300 K, what will be the initial susceptibility, χ_i? The Curie temperature of the material, T_c, is 700 K and γ is proportional to $\sqrt{(T_c - T)}$; what will be χ_i at 600 K? If the interatomic spacing is 1.5 Å, estimate the first magnetocrystalline anisotropy constant and the wall thickness at 300 K.

[Assume the J–T curve follows the tanh law.]

[Ans. 3200; 2500; 7.8 kJ/m³; 64 nm]

6.8. Figure P6.8 shows polarization curves for $PbFe_{12}O_{19}$ in the [00.1] (easy) and [10.0] (hard) directions. Assuming that only the first magnetocrystalline anisotropy constant (K_1) is non-zero, find it. What is the greatest possible value for $_jH_c$? The saturation polarization is found to be 0.4 T at 300 K, what is the greatest possible $(BH)_{max}$ at 300 K? What are the greatest possible values for B_r, H_c and $(BH)_{max}$ for a randomly oriented polycrystalline sample at 300 K?

[Ans. 226 kJ/m³; 1.13 MA/m; 31.8 kJ/m³; 0.2 T; 159 kA/m; 7.96 kJ/m³]

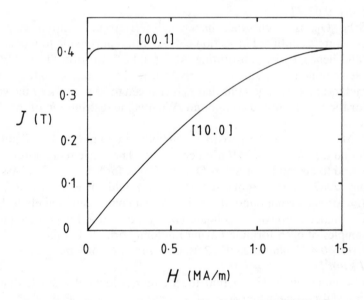

Figure P6.8   *Polarization anisotrophy in a single crystal of PbFe$_{12}$O$_{19}$*

6.9. A superparamagnetic material comprises very small ferromagnetic particles in a non-ferromagnetic matrix. Such materials have a polarization obeying the equation

$$j = L(x)$$

where $j = J/J_s$, J = measured polarization, J_s = saturation polarization of the particles, $x \equiv \mu H/k_B T$, μ is the particle moment ($= J_s V$, where V is the particle volume), H is the applied field and $L(x)$ is the Langevin function.

Show that $L(x) \approx \frac{1}{3}x$ if x is small. Using this approximation find from figure P6.9, which shows some experimental results for a stainless steel, the particle volume if $J_s = 0.5$ T.

[Ans. 2×10^{-28} m^3]

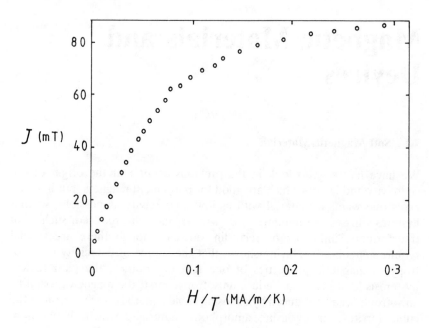

Figure P6.9   *J against H/T for a stainless steel sample*

7

Magnetic Materials and Devices

7.1 Soft Magnetic Materials

We have had a good look in the previous chapter at the origin of the coercivity and factors which are good for permanent magnets, but in many cases one wants a material with as low a coercivity as possible, so that hysteresis losses are minimized, or writing times are reduced, such as in transformers, inductors or recording media (though there are special requirements here). Having shown that the source of coercivity is anisotropy in magnetic properties of one sort or another, the path to low hysteresis losses is plain: reduce anisotropy. First, the magnetocrystalline anisotrophy can be reduced by choosing cubic materials, such as iron or the cubic ferrites, or by using amorphous materials, which, being noncrystalline, can have no magnetocrystalline anisotropy. Second, the particles should be large and multi-domain. And the domains should be free to move – no dispersed precipitates, voids or non-magnetic inclusions. Third, we must eliminate any strains by careful preparation and by choosing materials with low magnetostriction. A study of the preparation and properties of several different soft materials illustrates these principles.

7.1.1 Transformer Core Materials

Losses in transformer cores are made up of hysteresis losses, $\oint H dB$, eddy-current losses and anomalous losses. Hysteresis losses are obviously going to be a lot less than $4 H_c B_m$, where B_m is the maximum flux density in the core and H_c is the coercivity of the core. The only way to reduce these losses is to reduce H_c really.

Eddy currents flow in the same direction as the winding, but opposite in sense, so setting up an opposing field to that of the magnetizing current, causing a fall in flux density as the centre of the core is approached.

Lamination of the core so that the area of the laminations is normal to the eddy currents enables the flux to penetrate the core uniformly. The eddy-current losses in W/m^3 are given by

$$W_e = \rho J^2 = (\pi b f B_m)^2 \sigma / 6$$

where ρ is the resistivity, J is the eddy-current density, b is the thickness of the laminations, f is the frequency of excitation, B_m is the maximum flux density and σ is the conductivity. This expression for eddy-current losses is always too low, because of the anomalous losses arising from the non-ideal behaviour of domain walls in alternating fields. The anomalous losses are not easy to predict and vary from material to material according to the domain structure.

The important parameters for eddy-current losses, given f and B_m, are b and σ. The conductivity should be as low as possible, preferably zero, while the laminations should be as thin as convenient (but if $\sigma = 0$, there need be no lamination). Steel laminations are usually 0.3 mm thick but more recent materials are made of tape much thinner than this.

The most widely used material for large, low-frequency transformers is an alloy of iron and silicon, first used in 1905 and much improved since. Silicon reduces the electrical conductivity of iron, with little effect on the remanence. In 1934, Goss discovered a process which caused grain orientation in the material and reduced hysteresis losses. Goss material had losses at first of 2 W/kg, later reduced to 1 W/kg by secondary recrystallization.

The easy directions for magnetization in Fe and Fe–Si alloys are the <100> directions, which must therefore be oriented in the plane of the lamination. The Goss process achieves this, but is complicated, requiring a hot rolling, two cold rollings and an intermediate annealing at 900°C. Carbon is injurious and is removed by annealing in hydrogen at 800°C. Secondary recrystallization and grain growth are encouraged by a soak at 1200°C in pure, dry hydrogen. Large grain size helps to reduce hysteresis and resistivity losses due to grain boundaries. A texture known as 'cube on edge' is produced in which the easy <100> directions are along the rolling axis, in the plane of the laminations, while the <110> directions (not easy) are normal to this in the plane of the sheet as in figure 7.1(a). Now it is possible to produce a grain texture, known as 'cubex' having only <100> directions in the plane of the laminations as in figure 7.1(b).

Unfortunately, addition of more silicon to iron, though producing the desired benefit of lower conductivity, causes the iron to become brittle and hard to roll, though some effort has been made to produce an alloy with 6 per cent silicon, using vapour-phase deposition of silicon on Fe–3 per cent Si tape. Table 7.1 gives the relevant properties of iron and iron–silicon alloys, which shows that the addition of 3 per cent by weight of silicon decreases the electrical conductivity by a factor of five compared with

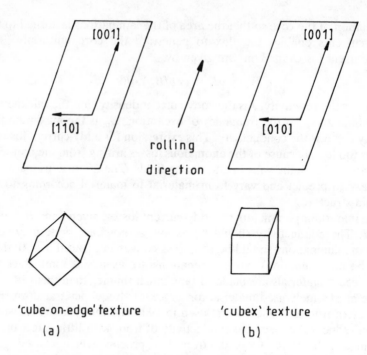

'cube-on-edge' texture

(a)

'cubex' texture

(b)

Figure 7.1 *(a) The 'cube-on-edge' texture in rolled silicon–steel sheet; (b) the 'cubex' texture in rolled silicon–steel sheet*

Table 7.1   Properties of Transformer Steels

Material*	σ**	Max. μ_r	H_c†	W_h‡	W_e‡	W_T‡
Commercial Fe	10	10^4	80	2.0	1.8	3.8
Fe–$\frac{1}{2}$%Si	15	5×10^3	70	1.8	1.2	3.0
Fe–3%Si	2	4×10^4	12	0.25	0.35	0.6
Fe–6%Si	1.3	—	8	0.2	0.2	0.4

* In the form of 0.3 mm thick laminations.
** Conductivity in MS/m.
† Coercivity in A/m.
‡ Losses in W/kg at 50 Hz and $B_m = 1$ T.

commercial soft iron, while the hysteresis losses are halved. Increasing the flux density from 1 T to 1.5 T more than doubles the total losses.

More recently, the inherent anisotropy of metallic glasses has been exploited. If a molten metal, containing stabilizers such as boron, is solidified very rapidly (rates of 10^4 K/s are required), the atoms are unable to form regular crystallites and remain in the relative positions they had while liquid, that is they are glassy. This disorder naturally causes reduced conductivity, while its randomness ensures magnetic anisotropy. Both eddy-current and hysteresis losses are thereby reduced. Table 7.2 gives data for some of the compositions, based chiefly on iron, with additions of boron and silicon, indicating that their losses are appreciably less than those of identical laminations made from silicon–iron.

At 50 or 60 Hz, core losses in large transformers (according to table 7.1) are about 600 W/tonne, so that if copper (winding, I^2R) losses are about the same, or a little more, the total losses would be 1.5 kW/tonne. Now a 1 tonne transformer core (the winding, casing, coolant, terminals etc. will weigh in addition rather more than this on their own) will have a rating of about 200 kVA, so the efficiency is very high at full load and a power factor of unity (99.25 per cent). However, the cost of core power losses alone even in a highly efficient transformer works out at £250/tonne/year. The core losses continue even though no load is supplied, of course.

Table 7.2   Properties of some Metallic Glasses for Transformers

Composition	J_s^a	H_c^b	σ^c	T_c^d	W_T^e
$Fe_{81}B_{13.5}Si_{3.5}C_2$	1.6	3.2	0.77	643	0.3
$Fe_{67}Co_{18}B_{14}Si_1$	1.8	4.0	0.77	688	0.5
$Fe_{40}Ni_{38}Mo_4B_{18}$	0.9	1.2	0.62	626	0.4
$Fe_{78}B_{13}Si_9$	1.5	1.5	0.8	600	0.3
$Fe–6\%Si^\dagger$	1.9	8.0	1.3	900	0.8

† For comparison.
^{a} J_s in T at 300 K.
^{b} Induction coercivity in A/m.
^{c} Conductivity in MS/m at 300 K.
^{d} Curie temperature in K.
^{e} Total losses in W/kg at $B_m = 1.4$ T.

7.1.2   Soft Ferrites

Eddy-current losses can be eliminated completely if the core is non-conducting, as in a ferrite. Before the advent of ferrites during World War II it was very hard to make inductor cores which worked well much above the audio range of frequencies, though some success was had with iron dust cores at up to 1 MHz (and today molypermalloy powder cores are made which are competitive with ferrites up to perhaps 300 kHz, though their maximum induction is only about 0.5 T). The wartime exploitation of radar had made the manufacture of non-conducting magnetic materials essential, though the first publication on soft ferrites did not emerge until afterwards (*Philips Technical Review*, **8**, 353 (1946)).

The soft ferrites, of general formula AB_2O_4 (or $AO.B_2O_3$, where A is a divalent metal and B is trivalent, usually Fe^{3+}) have a cubic crystal structure known as the *spinel* structure after the mineral spinel, $MgAl_2O_4$, which is not magnetic. The spinel ferrites have an interesting structure consisting of cubic close-packed O^{2-} ions with transition metal cations such as Ni^{2+} and Fe^{3+} in the interstitial sites, as in figure 7.2, which shows an eighth part of the unit cell in plan form. (Note that the *c*-distances marked on figure 7.2 are half those in the true unit cell.) The unit cell contains 8 formula units, 64 tetrahedral (four-fold coordinated) interstitial sites, marked X in figure 7.2, and 32 octahedral (six-fold coordinated) sites, marked • in figure 7.2. In a normal spinel, such as $ZnFe_2O_4$, the Zn^{2+} cations occupy 8 of the 64 tetrahedral sites, while the 16 Fe^{3+} ions occupy half of the octahedral sites. But zinc ferrite is antiferromagnetic, so half the Fe^{3+} ions have spins pointing up and half point down (they are said to belong to two different sublattices of opposing moments).

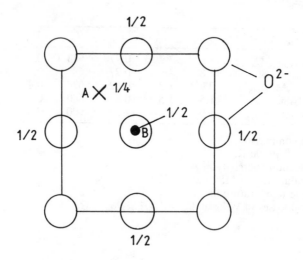

Figure 7.2 *An eighth part of the spinel unit cell*

In some cases the A^{2+} ions occupy the B-sites (the octahedral sites) and displace half the B^{3+} ions into tetrahedral sites. The resulting structure is known as the *inverse spinel* structure, and has a different magnetic order from that of a normal spinel. Suppose A is Ni^{2+} and B is Fe^{3+}; half the Fe^{3+} ions are on tetrahedral sites and these align antiparallel to *all* the ions in the octahedral sites:

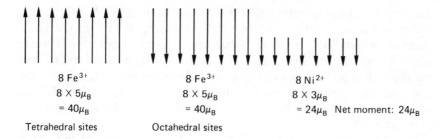

8 Fe^{3+}	8 Fe^{3+}	8 Ni^{2+}	
8 $\times$ 5μ_B	8 $\times$ 5μ_B	8 $\times$ 3μ_B	
= 40μ_B	= 40μ_B	= 24μ_B	Net moment: 24μ_B
Tetrahedral sites	Octahedral sites		

The resultant moment is only 24 μ_B. The formula can be written $[Fe^{3+}][Fe^{3+}Ni^{2+}]O_4$.

If the divalent cation is to have maximum magnetic moment it must have five unpaired electrons, Mn^{2+}, in fact. $MnFe_2O_4$ should have a saturation polarization (at 0 K) corresponding to 40μ_B per unit cell. Now the unit cell in all the spinels has a lattice parameter of about 0.8 nm (determined by the size of the oxygen anion – see question 7.2), so the magnetic moment/m^3 is 40 μ_B/a^3, which gives J_0 (if μ_B is in Tm^3) to be 0.91 T, whereas the value of J_s at 300 K is only 0.52 T, because the Curie temperature of $MnFe_2O_4$ is 573 K. Equation (6.4) suggests that for $t = 0.52$ (that is, $T = 298$ K), $J = 0.95$ (or $J_s = 0.86$ T), but as the material is ferrimagnetic, equation (6.4) does not apply. In ferrimagnets the opposed sublattice magnetizations individually follow a tanh law, or a Brillouin function which is similar, but their resultant is roughly a straight line with some bending at the ends, as in figure 6.5. A linear fall of j with t would give $j = 0.47$ (or $J_s = 0.43$ T) at 298 K – closer to the observed value.

Again, attempts were made to increase the rather low saturation polarization values of the ferrimagnets beyond what could be obtained in manganese ferrite. One way would be to increase T_c, so that J_s would be nearer to J_0, but it appears that this is impossible. In fact the highest Curie temperature in a spinel ferrite is found with $FeO.Fe_2O_3$ – magnetite, the oldest magnet of all – at 860 K. Unfortunately, magnetite has a lower saturation induction than $MnFe_2O_4$, as Fe^{2+} has a moment of 4μ_B, not 5μ_B as in Mn^{2+}. Another way of increasing J_s was tried in which Zn^{2+}, which always goes on the A sites, was substituted for Mn^{2+}. The zinc ions displaced an equal number of ferric ions from the A sites, so that the polarization was increased (see problem 7.1). However, when more than about half of the Mn^{2+} was replaced, the magnetic moment fell once more as the magnetic ordering was affected by an excess of zinc.

A material based on a framework of close-packed oxygen ions, such as the spinel structure, is capable of almost infinite modification, since many ions will fit into the framework without destroying the structure: all that is needed is to maintain charge neutrality overall and make sure the ions are the right size. For example, Fe^{3+} has an ionic radius of 0.7 Å and the ionic radius of Li^+ is about 0.7 Å, so it will substitute for Fe^{3+} in the Fe_3O_4 structure without distortion. However, the requirement of overall charge neutrality means that some of the Fe^{2+} ions must be oxidized to Fe^{3+} and not more than half may be replaced, as in the composition $Li^I_{0.5}Fe^{III}_{2.5}O_4$. One can also substitute tetravalent ions of the right size, such as Ti^{4+}, in which case some of the remaining iron must become divalent. (See also problem 7.4.)

Mixed ferrites such as $(Mn, Zn)Fe_2O_4$ are made, not so much to increase J_s, but to improve properties such as permeability and resistivity. The permeability may be increased by reducing the Curie temperature, as then the wall energy is reduced in accordance with equation (6.11), and some reduction in polarization is a consequence. Some of the properties of commercially made spinel ferrites are given in table 7.3. Appreciable electrical conductivity occurs in many ferrite compositions because of semiconductivity associated with holes in the $3d$ bands, though the hole mobility is rather low as the bands are narrow.

Table 7.3   Properties of some Spinel Ferrites

Formula	J_0	$J_s{}^a$	$\mu_i{}^b$	σ^c	$T_c(K)$
$MnFe_2O_4$	0.8	0.5	250	50	570
Fe_3O_4	0.7	0.6	70	2.5×10^4	860
$NiFe_2O_4$	0.4	0.34	10	10^{-6}	860
$CoFe_2O_4{}^d$	0.6	0.53	1	10^{-5}–10^{-1}	770
$CuFe_2O_4$	0.2	0.17	—	—	730
$MgFe_2O_4{}^e$	$\begin{cases} 0.22 \\ 0.56 \end{cases}$	0.12 / 0.40	—	—	578 / 710
$(Mn, Ni)Fe_2O_4{}^f$	0.6	0.22	500	10^{-3}	400
$(Mn, Zn)Fe_2O_4{}^g$	1.25	0.35	1200	1	420

[a] J_s at 300 K.
[b] Initial relative permeability.
[c] Electrical conductivity at 300 K in S/m.
[d] Not a soft ferrite, but a permanent magnet.
   σ varies widely as the Co/Fe ratio goes from 0.49 to 0.51.
[e] Propeties depend on degree of inversion.
[f] Commercial material, Ferroxcube™ B1.
[g] Commercial material, Ferroxcube™ A4.
   Ferroxcube is a trademark of the Mullard plc.

7.1.3   Production Technology of Soft Ferrites

Ceramic materials require special techniques for production and fabrication – techniques quite different from those of metallurgy. Ferrites used for transformer/inductor cores, as well as those used for permanent magnets, are not required in single-crystal form, so they are usually produced as powders by a batch process. The first requirement is to make the chemical compound with the precise formula required. Chemicals of high purity are used, as small amounts of impurities can degrade performance.

Suppose we want to make $(Ni, Zn)Fe_2O_4$. We must mix together compounds of the cations in the correct proportions which are then intimately mixed as fine powders and heated to react and give nickel–zinc ferrite. A convenient source of the iron might be ferrous oxalate, which is easily purified and decomposes to CO_2, H_2O and Fe_2O_3, in oxidizing conditions. Nickel could come from the acetate or carbonate, while zinc could be obtained from the carbonate also. Carbonates are favoured over the oxides because they are readily purified and react at lower temperatures.

The powdered reactants are usually mixed together and broken down into smaller particles by ball-milling. A ball mill consists of a rotatable steel drum with steel balls in it that smash up the brittle powders (this is called *comminution*). In some cases, such as when iron is an undesirable impurity, the ball mill must be lined with another material and different balls must be used. A liquid such as ethanol may be added to keep down the dust and assist comminution. Very fine powders are necessary because the solid-state reactions will only occur as a result of diffusion of the reactants, and the reaction temperature should be as low as possible, so that milling for 24 hours may be required.

Once mixed and comminuted, the powdered reactants are dried and fired to give the end product in the form of a sintered, porous, solid mass, which must be broken up once more in a ball mill to give a fine powder (typical particle size: ~ 1 μm). The progress of the solid-state reaction can be followed with the help of two techniques called thermogravimetric analysis (TGA) and differential thermal analysis (DTA). In both cases the reactants are heated at a constant rate, while in TGA the sample's weight is plotted as a function of temperature. As the reaction proceeds, gases are evolved and their evolution stops when it is complete. Figure 7.3(a) shows a TGA apparatus and figure 7.3(b) a TGA plot for a sample evolving both H_2O and CO_2. Once the ideal conditions for the reaction have been established – provided the reactants are prepared the same way – no further TGA plots are necessary and full-scale production may proceed.

A DTA apparatus is shown in figure 7.4(a). In DTA the temperature of a small sample of the powder to be reacted is compared with that of a similarly sized, inert material. The temperature is measured by thermocouples connected in series opposition and the difference voltage (which may be only 0.1 μV) is amplified and recorded as a function of time along with the reference temperature. Any heat evolved (or absorbed) by reactions in the sample results in an output voltage as shown in figure

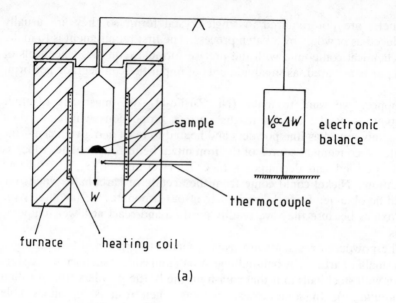

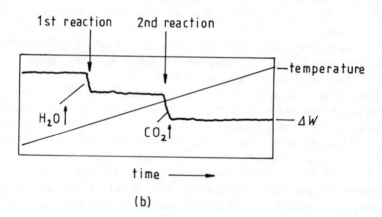

Figure 7.3 *(a) An apparatus for thermogravimetric analysis (TGA); (b) results of a TGA experiment*

7.4(b). When the reaction is complete, no heat is evolved from the sample and the DTA output is a straight line. As with TGA, once the reaction conditions are established, no further use need be made of DTA, unless a batch of reactant powder is suspect.

To go back to the manufacture of nickel–zinc ferrite, since the nickel is to remain in oxidation state II, the iron[II] oxalate and zinc carbonate are

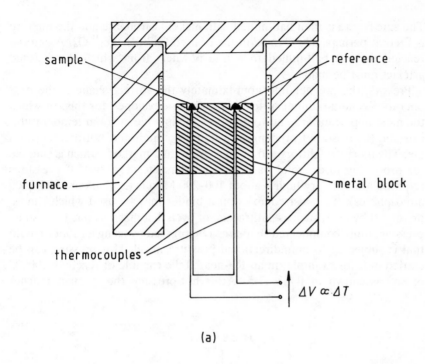

(a)

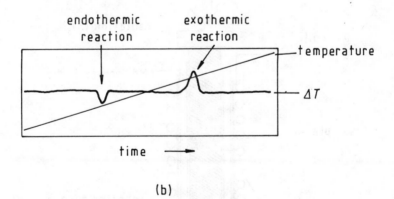

(b)

Figure 7.4 *(a) An apparatus for differential thermal analysis (DTA); (b) results from a DTA experiment*

first reacted together in an oxidizing atmosphere at about 600°C to form $ZnFe_2^{III}O_4$:

$$4Fe(COO)_2 + 2ZnCO_3 + 3O_2 \rightarrow 2ZnFe_2O_4 + 10CO_2 \uparrow$$

The zinc ferrite is then ball-milled with nickel[II] carbonate and the mixture is fired at perhaps 800°C to form the desired $(Ni^{II},Zn)Fe_2^{III}O_4$. Again the reacted mass is ball-milled to a fine powder, from which the finished artefact must be made.

Pressing the powder into approximately the correct shape is the next operation, and a number of techniques have developed for this, of which the most important are end-to-end pressing in a die at room temperature, isostatic (hydrostatic) pressing in a liquid medium and hot-pressing in a die. The most straightforward of these is the first method, which is suitable for producing discs, solid bars and cylinders etc. The sort of pressures required are not very high – about 100–200 MPa (1–2 kbars or 1000–2000 atmospheres). To assist compaction a binder may be used which can be driven off by heat, such as anthracene or even deionized water. In isostatic pressing, quite complicated shapes can be pressed by using a rubber mould that is subjected to omnidirectional pressure in oil. Hot pressing can be carried out, for example, in an RF-heated die capable of reaching 1000°C or so, as shown in figure 7.5. After hot-pressing the artefact reaches

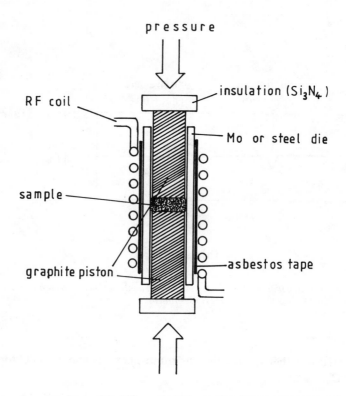

Figure 7.5 *An apparatus for hot-pressing powders*

100 per cent of the theoretical density, but after cold pressing, the 'green' artefact is porous ($\approx$ 60 per cent theoretical density) and must be handled with care to avoid damage.

After cold pressing, the green pieces must be fired ('sintered') at high temperatures, such as 1200°C, to achieve perhaps 80–90 per cent of theoretical density; substantial porosity remains. Sintering is a very critical process in the production of an artefact with the right properties; it must result in the proper grain size and density, while not causing unpredictable shrinkage. Dimensional control in sintering is not possible beyond $\pm$ 2 per cent, so the final shaping must be done by grinding. Compared with metal-forming, the making of ceramic artefacts seems difficult and complicated, yet it is an ancient craft rediscovered and developed to the point where quality and cheapness are not incompatible.

7.2 Materials in Magnetic Recording

In cost, magnetic recording is the most important use for magnetic materials today. Though semiconductors have displaced the ferrite ring, after a long struggle, from computer fast-access memories, magnetic bubbles have come along as a formidable competitor. The chief uses for magnetic materials in recording are

(1) Magnetic tape
(2) Floppy discs
(3) Hard discs
(4) Recording heads.

Until recently, all the first three have utilized γ-Fe_2O_3 or cobalt-modified γ-Fe_2O_3 as the recording medium and the recording has been horizontal, or transverse, as shown in figure 7.6(a), rather than perpendicular as in figure 7.6(b). The heads are all highly permeable magnetic rings of Permalloy (Ni–Fe), Alfesil (Al-Fe-Si) or a soft ferrite, with very small gaps (about 1 μm) to produce the fringing field that does the writing. Small gaps imply a small depth of fringing field and thus a small head-recording medium gap. The coercivity of the medium determines how stable the recording is, but cannot be too high for the fringing, writing field. As the gap between head and medium has decreased because of better technology in producing the recording media and in the mechanics of the recorder, so the permissible intrinsic coercivity has risen from 2.4 kA/m (γ-Fe_2O_3) to 3.6 kA/m (CrO_2) to 4.8 kA/m (cobalt-modified γ-Fe_2O_3) to 10 kA/m (oxidized Fe). Hard magnetic materials such as the hexaferrites are now finding favour for perpendicular media.

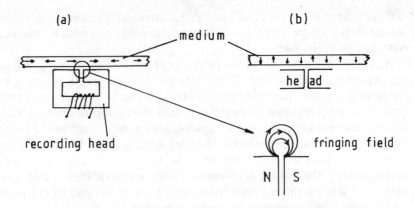

Figure 7.6   *(a) A horizontal recording medium; (b) a perpendicular recording medium*

The γ-Fe$_2$O$_3$ particles used in recording media are about 0.5×0.1 μm in size, with substantial porosity. The chemical process for their production is fairly complicated:

(1) $FeSO_4 + NH_4OH \longrightarrow Fe(OH)_2 \downarrow$

(2) $Fe(OH)_2 \xrightarrow{O_2} \alpha$-Fe$_2O_3$.H$_2$O (goethite, yellow oxide)

(3) α-Fe$_2$O$_3$.H$_2$O $\xrightarrow{200^{\circ}C} \alpha$-Fe$_2O_3$ (hex-R, red oxide)

(4) α-Fe$_2$O$_3 \xrightarrow{H_2}$ Fe$_3$O$_4$ (magnetite)

(5) $Fe_3O_4 \xrightarrow[O_2]{200^{\circ}C} \gamma$-Fe$_2O_3$ (fcc, brown oxide)

Magnetite produced in this way is not stable, so it is further oxidized to metastable γ-Fe$_2$O$_3$, which is ferrimagnetic also. The moment of γ-Fe$_2$O$_3$ is about 2.5 μ_B per formula unit, compared with 4 μ_B for magnetite, and it may best be thought of as being a defect spinel in which one-sixth of the B-site ions are missing, while all the iron ions are trivalent, that is $[Fe^{III}][Fe^{III}_{5/3}]O_4$. Heating γ-oxide to 400°C turns it to α-oxide, which is the common, non-magnetic, rhombohedral form known as haematite. The fine particles of γ-Fe$_2$O$_3$ are made into a slurry with a polyurethane resin and this is used to coat mylar film at high speed (2 m/s). After coating, the film is subjected to a magnetic field of about 80 kA/m to orientate the particles. This improves the squareness of the hysteresis loop, and reduces the coercivity range (important for uniform recording) as might be expected for single-domain particles from Stoner–Wohlfarth theory. Squareness factors of 0.8 are usual.

Floppy discs are produced by the same process as tapes, but hard discs require different processing. For these, an aluminium disc is oxidized to give a surface layer of alumina, onto which is sprayed an epoxy-based slurry of half the particle density/m^3 as that used for mylar tape. After orienting, the slurry is dried, hardened and then polished to give a very smooth surface.

The coercivity of pure γ-oxide films is rather low at 2.4 kA/m and so CrO_2 was used for a time to improve this; however CrO_2 had to be produced by an expensive process and was soon replaced by cobalt-modified γ-Fe_2O_3. A very thin layer of cobalt hydroxide is produced on the surface of the γ-Fe_2O_3 particles by placing them in a solution of a cobaltous salt to which sodium hydroxide is added. The particles are removed and dried at 150°C to form a surface layer of $CoFe_2O_4$, so that the film has an intrinsic coercivity double that of unmodified γ-oxide. Next to be tried were films made from reduced γ-Fe_2O_3 which were in effect just iron. The aspect ratio of the particles (about 5 : 1) leads to considerable coercivity from shape anisotropy, and figures of 10 kA/m are quoted. Progress is now well advanced for producing perpendicular media from $BaFe_{12}O_{19}$ particles.

An important consideration for magnetic recording media is the particle density, which was about 10^{20}/m^3 for γ-Fe_2O_3, rising to 20^{21}/m^3 for cobalt-modified γ-oxide and to 10^{22}/m^3 for reduced γ-oxide. Suppose the depth of the oxide coating is 1 μm, the trackwidth is 0.25 mm and the width of the write-field is 2 μm, then there are going to be about 50 000 magnetized particles to read in a γ-Fe_2O_3 film, which is sufficient. But the entire film volume in a hard disc may be only 5×10^{-9}/m^3, of which about 20 per cent is used, to give 10^{-9}/m^3 as the volume written to. If it is to contain 20 Mbytes or 120 Mbits, then each bit occupies only about 10^{-17} m^3, so a γ-Fe_2O_3 medium with 10^{20} particles/m^3 would have only about 1000 particles/bit. This would not be acceptable as such a small number of particles could not reliably retain the information. Greater disc capacity can only be obtained with finer particles.

7.3 Magnetic Bubbles

Magnetic bubbles have made some headway against the tide of semiconductors in memory applications. A magnetic bubble is merely a cylindrical domain of reversed polarization which can be formed in a thin platelet of ferromagnetic material and which can be used to store information. By forming suitable electrodes on the material, the bubbles can be moved from one storage location to another. Permalloy (a highly permeable alloy of nickel and iron) has been used as an electrode material. Bubbles can be detected by the stray field at their circumferences using, for example, magnetoresistive sensors. Since the film containing the bubbles is very thin, its demagnetizing field is J_s/μ_0, which must be less than the magneto-crystalline anisotropy field, $2K_1/J_s$, if the magnetization is to be directed normal to the plate, that is

$$2K_1/J_s > J_s/\mu_0$$

or

$$K_1 > J_s^2/2\mu_0 \tag{7.1}$$

which is the primary condition for bubble formation.

A thin plate of suitable bubble material will spontaneously form domains of opposed magnetization as in figure 7.7(a), where they are in the shape of meandering stripes. The magnetization direction is along the easy axis, perpendicular to the platelet. The areas of opposed polarization are about equal in the absence of an applied field. When a field is applied perpendicular to the plane of the plate, oppositely polarized domains shrink and eventually form a small cylindrical domain that is called a bubble, as in figure 7.7(b). The diameter of the bubble must be as small as possible to give maximum bit density: typically about $10^{10}/m^2$ for 2 μm diameter bubbles. The length parameter, λ, which determines the bubble diameter is given by

$$\lambda = \mu_0\gamma/J_s^2 \tag{7.2}$$

Bubble diameters are about 8λ, when the film thickness is 4λ. Equation (7.2) shows that J_s must be large, and wall energy small for small bubbles. As condition (7.1) must be met, it means that K_1 should also be large, tending to increase wall energy.

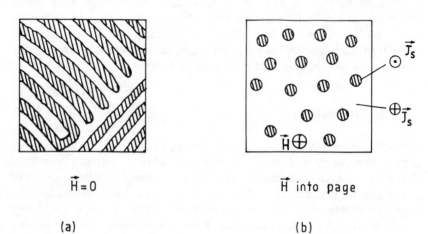

$\vec{H} = 0$

(a)

$\vec{H}$ into page

(b)

Figure 7.7 *(a) Stripe domains in a thin plate in zero field; (b) bubble domains which form in a thin plate when $H > H_{crit}$*

Bubble materials must have certain dynamic characteristics as well as the static properties of equations (7.1) and (7.2). The coercivity of the bubbles affects how fast they may be moved and generally it should meet the condition

$$\mu_0 H_c < 0.05 J_s$$

A class of ferrimagnets with the garnet structure are popular for making bubble devices since they have suitable values of J_s and K_1. Garnets have the general formula $A_3B_5O_{12}$, where both A and B are trivalent, and numerous compositions with A being a combination of rare earths or yttrium and B being a mixture of scandium, gallium or aluminium have been tried. The garnet structure is cubic and the ferrimagnetic garnets have a lattice parameter of about 12.4 Å. Epitaxial films of the ferrimagnet material (about 1–2 μm thick) are grown on a paramagnetic substrate which must be closely matched in lattice parameter. Since Ga^{3+} has almost the same ionic radius as Fe^{3+}, the most common substrate is gadolinium gallium garnet, $Gd_3Ga_5O_{12}$, or GGG. GGG can be grown by the Czochralski technique as single crystals up to 100 mm in diameter, while the epitaxial film is grown by CVD or liquid-phase epitaxy.

The thrust of the research effort has been to increase J_s to reduce bubble size and at the same time increase K_1 proportionately to obey (7.1), while keeping the coercivity below $0.05 J_s/\mu_0$. Bubble sizes of about 1 μm diameter have been achieved with velocities as high as 500 m/s in fields of 150 A/m (a mobility of 3.3 m^2/As). In a 10 mm by 10 mm 1 Mbit chip of Japanese design the bubble film is made from $(Y, Sm, Lu, Ca)_3(Ge, Fe)_5O_{12}$, having $J_s = 0.05$ T, with a bubble diameter and film thickness both about 2 μm. Further development will certainly be able to improve performance by ten-fold, which should keep bubble devices competitive with semiconductor devices.

7.4 Microwave Devices

The advent of insulating ferrimagnets opened up a new field for magnetic materials in the microwave region (roughly 300 MHz to 300 GHz, or from 1 m down to 1 mm in wavelength), which had been inaccessible to conducting materials. Many of these microwave devices are based on the principles of Faraday rotation and ferromagnetic resonance. The garnets are preferable to the ferrites for this application as they are of much higher resistivity.

7.4.1 The Isolator

A circularly polarized electromagnetic wave is one in which the electric and magnetic field vector (normal to the direction of propagation) rotates at a steady rate either in a clockwise sense or in an anticlockwise sense. Circularly polarized electromagnetic waves are subject to different attenuations and propagation velocities in a ferromagnet according to the sense of the circular polarization. Now a plane-polarized wave can be constructed from two identical, circularly polarized waves of opposite sense, so that if it propagates down a ferromagnetic rod the plane of polarization is rotated as shown in figure 7.8. In operation a biasing field is applied to the rod to saturate it.

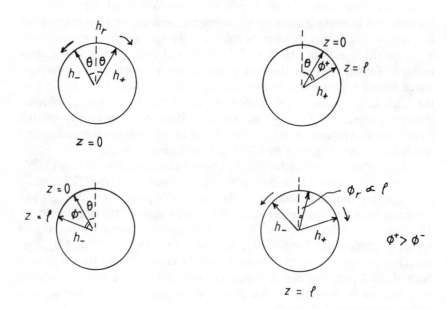

Figure 7.8   *The rotation of the polarization during propagation down a ferrimagnetic rod (Faraday rotation)*

Consider a plane-polarized wave travelling along a ferromagnetic rod in the direction of the biasing field ($+ z$). Clockwise and anticlockwise components of the wave propagate at different velocities, and this results in differing phase changes per unit length of rod. The rod length can be adjusted so that the phase change is exactly 45° when the wave travels through the whole of the rod. On the output side of the rod the waveguide is arranged at 45°C to the input waveguide, so that the forward wave is accepted. Any reflected wave will also be subjected to a phase shift of 45° in its reverse passage ($- z$ direction), but in the same sense as the original wave. The overall rotation of the reflected wave is thus 90° compared with the phase of the wave incident on the rod. Figure 7.9 shows the situation. The incoming waveguide attenuates waves which are 90° out of phase by 30 dB typically, so the isolator acts like a one-way valve.

The linewidth of the ferromagnet is a measure of its performance as an isolator. The linewidth is the width of the imaginary (absorbing) part of the plot of complex clockwise permeability against magnetic field. A small linewidth results in greater attenuation of unwanted radiation, and garnet materials have linewidths less than a third of that of the better ferrites. Tanδ losses (of which more in the next chapter) are also much less in the garnets because of their higher resistivities.

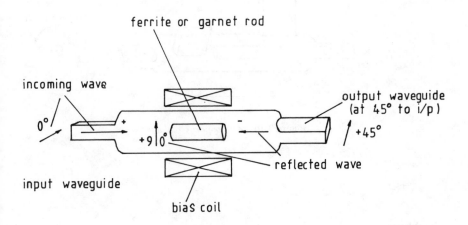

Figure 7.9   *The microwave isolator*

7.4.2 *Circulators*

A circulator is a microwave device having at least three ports, which are arranged so that the input to the *n*th port appears at the output of the (*n* + 1)th port and is zero elsewhere. Figure 7.10 shows a 4-port circulator arranged so that there is a 45° rotation going from port to port in a clockwise direction and therefore a 90° rotation going from port to port in the anticlockwise sense. Waveguide no. 2 is rotated 45° about the *z*-axis, so that the signal from the aerial (port 1) is accurately aligned to the receiver by the Faraday rotation in the ferrite. The transmitter waveguide is at 90° to the receiver, so that the latter picks up little energy from the former. The dummy load is to absorb any signal coming from the receiver.

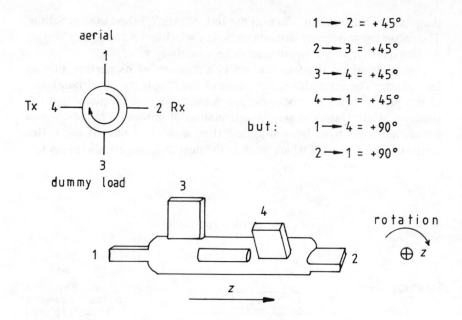

Figure 7.10 *A four-port circulator*

Problems

7.1. The Zn^{2+} ions in $Cu_{0.7}Zn_{0.3}Fe_2O_4$ have a strong preference for the tetrahedral (A) sites, so they displace equal numbers of Fe^{3+} ions from these sites to the octahedral (B) sites. The formula can be written $[Fe_{0.7}{}^{3+}Zn_{0.3}{}^{2+}][Cu_{0.7}{}^{2+}Fe_{1.3}{}^{3+}]O_4$, where the first bracket encloses the A-ions and the second the B-ions. If the A-ions align antiparallel to the B-ions, what is J_0 for $Cu_{0.7}Zn_{0.3}Fe_2O_4$? If $T_c = 400$ K for this compound, what is J_s at 300 K if dJ/dT is linear from 0 K to T_c? What would J_s be at 300 K if the polarization obeyed the tanh law?
[In practice, J_s is about 0.2 T at 300 K.]
[Ans. 0.67 T, 0.17 T, 0.52 T]

7.2. Lanthanum ferrite ($LaFeO_3$) has been used for magnetic bubble devices. Assuming that its crystal structure is cubic and comprises close-packed O^{2-} and La^{3+} ions with Fe^{3+} in appropriate interstitial sites, and given that the ionic radii of O^{2-}, Fe^{3+} and La^{3+} are 1.40 Å, 0.55 Å and 1.06 Å respectively, suggest a unit cell for $LaFeO_3$. Estimate its lattice parameter.

$LaFeO_3$ is a canted spin ferrimagnet in which half the Fe^{3+} ions have a polarization almost exactly opposite to that of the other half.
If J_0 is 0.01 T, find the angle between the moments.
[Ans. 169°]

7.3. A metal glass has a coercivity of 2 A/m and a maximum induction of 0.4 T when used in a 50 kHz transformer. Estimate the hysteresis loss/kg if the density is 7000 kg/m³ and the loop is square.
Eddy-current losses at this frequency are found to be four times the hysteresis losses. The total losses in W/kg are given by

$$W_T = af + bf^2$$

where f is the frequency in Hz, a is the hysteresis loss coefficient in J/kg and b is the eddy-current loss coefficient in Js/kg. If a ferrite material has $a = 2.85$ mJ/kg, above what frequency is its use advantageous?
[Assume infinite resistivity for the ferrite.]
[Ans. 22.9 W/kg, 65.4 kHz]

7.4. A ferrite with the spinel structure is made with iron in oxidation state III only and titanium in oxidation state IV only, and in a 1 : 1 atomic ratio. Give several ways of achieving overall charge neutrality in this material.

Calculate the value of J_0 for CrO_2, assuming the oxygen ions are close-packed and of radius 1.40 Å and that all the moments of the chromium ions are parallel. If J_s is 0.6 T at 280 K, estimate its Curie Temperature by (a) assuming that the magnetization obeys a tanh law and (b) that it falls linearly with temperature.
[Experimentally, T_c is 130°C.]
[Ans. 121°C, 1283°C]

7.5. Barium hexaferrite has a domain wall energy of about 6 mJ/m², and at 300 K, $J_s = 0.5$ T, while $K_1 = 380$ kJ/m³. Will magnetic bubbles form in this material? What will be the length parameter? Why is $BaFe_{12}O_{19}$ not used for bubble devices?
[Ans. 30 nm]

7.6. The relaxation time in seconds for the polarization of a particle of $\gamma\text{-}Fe_2O_3$ is given by

$$\tau = 10^{-9}\exp(J_sH_cV/2k_BT)$$

The particle's magnetization decays exponentially with this time constant after a saturating field is switched off.
If $H_c = 2.5$ kA/m, $J_s = 0.4$ T and $T = 300$ K, find the particle volume for $\tau = 1$ year and $\tau = 100$ years.
[Ans. $3.15 \times 10^{-22}m^3$, $3.5 \times 10^{-22}m^3$]

7.7. A molypermalloy powder material has core losses given by

$$W_T = aB_mf + b^2 + cf$$

where W_T is in W/kg, $a = 16$ mJ/kg/T, $b = 50$ nJs/kg and $c = 0.2$ mJs/kg; f is in Hz. If $B_m = 0.5$ T, at what frequency will the eddy-current losses equal the other losses? What will W_T be at this frequency? Would you use molypermalloy at this frequency? Compare the losses with those of the ferrite in question 7.3.

[Ans. 164 kHz]

7.8. The r.m.s. e.m.f. of a transformer winding is given by

$$E = 4.44 \, B_m A N f$$

where B_m is the maximum flux density in the core, A is the cross-sectional area of the core, N is the number of turns on the winding and f is the frequency. B_m is given by

$$B_m = \mu N I / l$$

where μ is the permeability of the core ($\equiv \mu_r \mu_0$), I is the peak current in the winding and l is the magnetic path length in the core. Show that for a given power rating, the volume of the transformer core is proportional to $1/f$.

What will be the weight of a ferrite transformer core rated at 500 VA at 20 kHz, if $B_m = 0.4$ T, $\mu_r = 500$ and its density is 5500 kg/m^3?

Why is the main electricity supply's frequency not higher than 50 Hz?

[Ans. 172 g]

8

Dielectrics

8.1 Electric Polarization

An ideal insulator contains no free electrons so that when an electric field is applied to it there is no macroscopic movement of charge; instead, it suffers displacement of its electrons with respect to their parental nuclei, creating thus electric dipoles. The electric field is said to (electrically) polarize the material. The dipole moment of the polarized atom in figure 8.1 is given by $\mu = q\xi$, where ξ is the displacement of the charge cloud. μ is directed from negative charge centre to positive charge centre. The units of dipole moment are thus Coulomb-metres (Cm) in SI units, but frequently used is the more conveniently sized unit, the Debye (D), which is 3.33×10^{-30} Cm (or 10^{-18} esu-cm). The polarization, P, is the dipole moment per unit volume (and thus has units of C/m^2):

$$P = N_v\mu$$

where N_v is the number of dipoles/m^3.

8.2 The Dielectric Constant or Relative Permittivity

Consider a parallel-plate capacitor as in figure 8.2(a) which has a plate area, A, plate separation, l, and carries a charge of $+ Q$ on the positive plate and $- Q$ on the negative. The surface charge density is given by Gauss's Law:

$$Q/A = \epsilon_0\mathscr{E} = \epsilon_0 V/l$$

where $\mathscr{E}$ is the field between the plates in a vacuum. Suppose now a slab of dielectric (polarizable material) is introduced between the plates so as to fill the space between them completely, as in figure 8.2(b). The electric field polarizes the dielectric and induces surface charge on it which must be

207

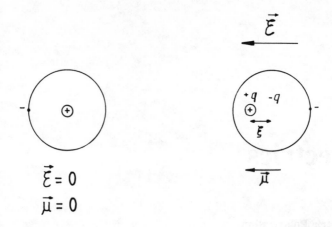

Figure 8.1   *The electric dipole moment induced on an atom by an applied electric field*

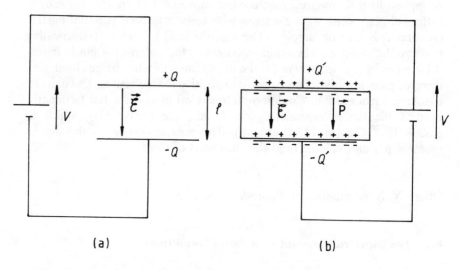

(a)   (b)

Figure 8.2   *(a) A parallel-plate capacitor with no dielectric; (b) a parallel-plate capacitor with a dielectric*

neutralized by extra charge on the plates of the capacitor. The new charge density is

$$Q'/A = \epsilon_0 \epsilon_r V/l = \epsilon \mathscr{E}$$

where ϵ_r is the relative permittivity, or dielectric constant, of the material. In the egs-emu system the permittivity of a vacuum is unity, so the factor ϵ_0 did not enter into equations, and the permittivity, ϵ, was the same as the relative permittivity, ϵ_r. The permittivity in the SI system is $\epsilon_r\epsilon_0$, often called the *absolute permittivity* to avoid confusion with ϵ_r. The electric displacement, D, is $\epsilon\mathscr{E}$ (compare $B = \mu H$) and the dielectric susceptibility, χ, is $\epsilon_r - 1$, so that

$$D = \epsilon_0(1 + \chi)\mathscr{E}$$
$$= \epsilon_0\mathscr{E} + \epsilon_0\chi\mathscr{E}$$

or

$$D = \epsilon_0\mathscr{E} + P \tag{8.1}$$

where $P \equiv \epsilon_0\chi\mathscr{E}$. Equation (8.1) is the electrostatic analogue of $B = \mu_0 H + J$ in magnetism.

The 'applied field' when the dielectric is present is $\mathscr{E}_0$ ($= D/\epsilon_0 = \mathscr{E} + P/\epsilon_0$); it has increased over the free space value by P/ϵ_0. The polarization is $(Q/A - Q'/A)$, the increase in capacitor charge per unit area. This definition of polarization can be shown to be equivalent to $P = N_v\mu$ by considering figure 8.3. If the area occupied by a dipole is δA, then the volume of a dipole is $l\delta A$ and so $N_v = 1/l\delta A$. The increase in charge density is $\Delta Q/A = q/\delta A$, where q is the charge on a dipole. Hence

$$P = \Delta Q/A = q/\delta A = qN_v l = N_v\mu$$

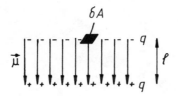

Figure 8.3 *Alignment of dipoles in the dielectric in a capacitor*

8.3 Types of Polarization

There are three principal classes of polarization in materials, illustrated in figure 8.4:

(1) Electronic Polarization: the displacement of orbiting electrons about an atom by an electric field.
(2) Molecular or Ionic Polarization: the displacement of ions by the electric field.
(3) Orientational Polarization: the partial alignment of polar molecules free to rotate in the electric field.

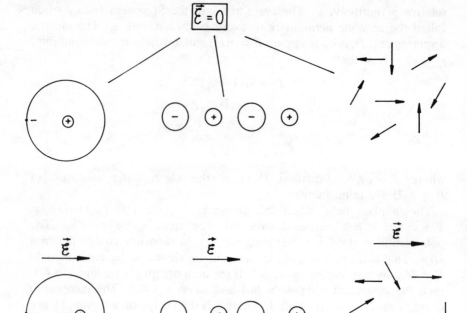

Figure 8.4   *The three types of electric polarization*

8.3.1   *Electronic Polarization*

The polarizability of an atom, α, is given by

$$\mu = \alpha \mathscr{E}$$

where $\mathscr{E}$ is the field at the site of the atom. For an atom of radius, r, the electronic polarizability is given by

$$\alpha = 4\pi\epsilon_0 r^3 \tag{8.2}$$

For instance in the hydrogen atom, $r = 0.5$ Å, so that $\alpha_H = 1.4 \times 10^{-41}$ F m^2, and in a field of 1 MV/m, $\mu = 1.4 \times 10^{-35}$ Cm. In solid paraffin, the density of hydrogen atoms might be about $10^{29}/m^3$, giving a (relatively small) polarization of 1.4×10^{-6} C/m^2. Equation (8.2) shows that large atoms are polarized much more than small.

8.3.2 Orientational Polarization

Polar molecules in liquids or gases (and even some solids), such as H_2O, CH_3Cl or NH_3, have a permanent dipole moment which can be partially aligned with an applied field. The measurement of the dipole moment of a molecule, allied with knowledge of its size, gives a measure of the charge separation in it which can be related to the atomic *electronegativities* (an index of how readily atoms will donate or accept electrons). Alignment of polar molecules in an electric field is upset by random thermal motion, so that Maxwell–Boltzmann statistics can be used to find the average polarization, just as for a paramagnet.

Consider a dipole whose moment, μ, makes an angle, ϑ, with the applied field, $\mathscr{E}$. Its potential energy is given by

$$U = -\mu.\mathscr{E} = -\mu\mathscr{E}\cos\vartheta$$

In a solid angle, $d\Omega$, there are $N_v d\Omega$ dipoles, where N_v is the number of dipoles per unit volume, given by

$$N_v = A\exp(-U/k_BT)$$

where A is a constant. The component of the dipole moment aligned with the field is $\mu\cos\vartheta$, so that the net dipole moment of solid angle, $d\Omega$, is $N_v\mu\cos\vartheta d\Omega$, and the mean dipole moment is

$$<\mu> = \frac{\int_0^\pi N_v\mu\cos\vartheta \, d\Omega}{\int_0^\pi N_v d\Omega} = \frac{\int_0^\pi N_v\mu\cos\vartheta.2\pi\sin\vartheta d\vartheta}{\int_0^\pi N_v.2\pi\sin\vartheta \, d\vartheta}$$

as $d\Omega = 2\pi\sin\vartheta d\vartheta$. Substituting for N_v leads to

$$<\mu> = \frac{\mu\int_0^\pi \exp(\mu\mathscr{E}\cos\vartheta/k_BT)\sin\vartheta \cos\vartheta \, d\vartheta}{\int_0^\pi \exp(\mu\mathscr{E}\cos\vartheta/k_BT)\sin\vartheta \, d\vartheta}$$

Making the substitutions $a \equiv \mu\mathscr{E}/k_BT$ and $x \equiv \cos\vartheta$ gives

$$<\mu> = \frac{\mu\int_{-1}^{+1} x\exp(ax)dx}{\int_{-1}^{+1}\exp(ax)dx} = \mu[\coth(a) - 1/a]$$

The integrals are standard forms: $\coth(a) - 1/a$ is the Langevin function, $L(a)$, which has been used in chapter 6.

When the argument, a, of $L(a)$ is small, $L(a) \approx \frac{1}{3}a$ (see problem 6.9), and then $<\mu> \approx \frac{1}{3}\mu^2\mathscr{E}/k_BT$ and the polarization is

$$P = N_v<\mu> \approx \frac{1}{3}N_v\mu^2\mathscr{E}/k_BT \tag{8.3}$$

To check if $\mu\mathscr{E}/k_BT$ is small, consider water, which has a dipole moment of 1.8 D or 6×10^{-30} Cm, to which a field of 1 MV/m is applied at 300 K. Then $\mu\mathscr{E} = 6 \times 10^{-24}$ J and $k_BT = 4 \times 10^{-21}$ J, so (8.3) holds true under normal circumstances.

8.3.3   The Total Polarization

The total polarization of a material is the sum of electronic, molecular and orientational polarizations:

$$P_{tot} = P_{el} + P_{mol} + P_{dipole}$$

The first two terms are almost independent of temperature. Using $P = N_v\mu$ and $\mu = \mathscr{E}_{loc}$ gives

$$P_{tot} = N_v\mathscr{E}_{loc}(\alpha_{el} + \alpha_{mol} + \alpha_{dipole})$$

$$= N_v\mathscr{E}_{loc}(\alpha_{el} + \alpha_{mol} + \mu^2/3k_BT)$$

$\mathscr{E}_{loc}$ is the field acting locally on the dipole, and is in general different from the applied field, $\mathscr{E}$. Again taking water as an example, the dipolar polarizability is about 3×10^{-39} Fm^2 at 300 K and the electronic polarizability $(4\pi\epsilon_0 r^3)$ is about 10^{-40} Fm^2, taking $r = 1$ Å. The dipolar term dominates and there is no molecular polarizability.

8.4   The Local Field in a Dielectric

The local field on an atom in a dielectric is made up of four terms:

$$\mathscr{E}_{loc} = \mathscr{E}_0 + \mathscr{E}_1 + \mathscr{E}_2 + \mathscr{E}_3$$

$\mathscr{E}_0(= D/\epsilon_0)$ is the external, or applied, field. $\mathscr{E}_1$ is the depolarization field arising from the charge induced on the surface of the dielectric, as in figure 8.5. $\mathscr{E}_2$ is the Lorentz cavity field which is produced by the surface charge on an (imaginary) spherical cavity centred on the reference atom (see figure 8.5). $\mathscr{E}_3$ is the field arising from atoms on crystal sites within the Lorentzian cavity, and is zero for cubic crystals and isotropic, amorphous materials. We shall consider $\mathscr{E}_3$ always to be zero.

$\mathscr{E}_1$, the depolarization field, arises in exactly the same way as the magnetic depolarization field, so that

$$\mathscr{E}_1 = -N_dP$$

The depolarization factor, N_d, is $1/\epsilon_0$ for a thin slab held normal to $\mathscr{E}$, zero for a needle with $\mathscr{E}$ along its axis and $1/3\epsilon_0$ for a sphere.

The Lorentz field, $\mathscr{E}_2$, acts in the same direction as P, as seen in figure 8.5, since the charges on the surface of the cavity must be the opposite way round to those on the external surface of the sample. Lorentz first

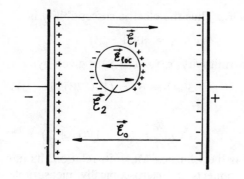

Figure 8.5 *Fields in a dielectric with a cavity*

calculated $\mathscr{E}_2$ and found it to be $P/3\epsilon_0$. Considering the slab of dielectric in figure 8.5 to be very thin, and taking $\mathscr{E}_3$ as zero, we find the local field to be

$$\mathscr{E}_{\text{loc}} = \mathscr{E}_0 - P/\epsilon_0 + P/3\epsilon_0$$

$$= \mathscr{E}_0 - 2P/3\epsilon_0$$

If this slab of dielectric is placed between the plates of a capacitor, the applied field, $\mathscr{E}_0$, from equation (8.1) is $\mathscr{E} = P/\epsilon_0$, so that

$$\mathscr{E}_{\text{loc}} = \mathscr{E} + P/\epsilon_0 - 2P/3\epsilon_0$$

or

$$\mathscr{E}_{\text{loc}} = \mathscr{E} + P/3\epsilon_0 \qquad (8.4)$$

Equation (8.4) is known as the *Lorentz relation*, from which important results follow, the most interesting being the *Clausius–Mossotti relation*.

8.5 The Clausius–Mossotti Relation

In the dielectric we know that

$$P = N_v\mu = N_v\alpha\mathscr{E}_{\text{loc}}$$

so that

$$\chi = P/\epsilon_0\mathscr{E} = N_v\alpha\mathscr{E}_{\text{loc}}/\epsilon_0\mathscr{E}$$

Substituting for $\mathscr{E}_{\text{loc}}$ from (8.4) gives

$$\chi = (N_v\alpha/\epsilon_0\mathscr{E})(\mathscr{E} + P/3\epsilon_0)$$

$$= 3\beta + \beta P/\epsilon_0\mathscr{E}$$

$$= 3\beta + \beta\chi$$

writing β for $N_v\alpha/3\epsilon_0$. Thus the electric susceptibility is

$$\chi = 3\beta/(1 - \beta) \tag{8.5}$$

and the relative permittivity, ϵ_r ($\equiv 1 + \chi$) is given by

$$\epsilon_r = 1 + 3\beta/(1 - \beta) = (1 + 2\beta)/(1 - \beta)$$

which gives

$$\beta = \alpha N_v/3\epsilon_0 = (\epsilon_r - 1)/(\epsilon_r + 2) \tag{8.6}$$

Equation (8.6) is the Clausius–Mossotti relation; its importance lies in the fact that it connects a macroscopically measurable quantity, the dielectric constant (ϵ_r), with an atomic property, the polarizability, α. If several sorts of atom are present in a solid, then αN_v in (8.6) becomes $\Sigma \alpha N_v$ for all the atoms. When ϵ_r is replaced by n^2 in (8.6), where n is the refractive index (see section 8.5.1), the equation is called the Lorenz–Lorentz relation, discovered in 1880.

8.5.1   The Refractive Index

The refractive index of a material (n) is the ratio of the speed of light in a vacuum (c) to the speed of light in the material (v). In chapter 2 we saw that

$$c = 1/\sqrt{(\mu_0\epsilon_0)}$$

while

$$v = 1/\sqrt{(\mu\epsilon)} = 1/\sqrt{(\mu_r\mu_0\epsilon_r\epsilon_0)}$$

so that

$$n = c/v = \sqrt{(\mu_r\epsilon_r)}$$

And in most optical materials and dielectrics $\mu_r = 1$, giving the very simple relationship

$$n = \sqrt{\epsilon_r}$$

Because optical frequencies are very high, only the electronic polarization can contribute to the dielectric constant in this region. As a result, ϵ_r in the optical region (or ϵ_∞, as it can be termed) is generally lower than the static value, ϵ_s. This is discussed in section 8.6.1.

We can make use of the Clausius–Mossotti relation to predict refractive indices: all we need to know are the atomic concentrations and electronic polarizabilities in the solid and use the equation above for the refractive index. As an example of this consider NaCl, which is cubic, has a lattice parameter of 5.64 Å and four formula units per unit cell, so that N_v is $4/(5.64 \times 10^{-10})^3$, or $2.23 \times 10^{28}/m^3$ for both Na^+ and Cl^- ions. From table 8.1 we find the polarizabilities as 0.16×10^{-40} and 3.30×10^{-40} Fm^2 so

$$\Sigma \alpha N_v = 2.23 \times 10^{28} \times (0.16 + 3.30) \times 10^{-40}$$

$$= 7.72 \times 10^{-12} \text{ F/m}$$

Substituting in (8.6) yields

$$0.291 = (\epsilon_r - 1)/(\epsilon_r + 2)$$

so that $\epsilon_r = 2.23$ and $n = 1.49$. The measured value is 1.50. Table 8.1 was derived from measurements of the refractive indices of ionic compounds, so it should be expected to lead to the right value for sodium chloride.

Table 8.1 Electronic Polarizabilities of Ions ($\times 10^{-40}$ Fm^2)

Ion	Al^{3+}	Be^{2+}	Ba^{2+}	Br^-	Ca^{2+}	Cl^-	Cs^+	F^-	I^-
α	0.06	0.01	2.8	4.63	1.22	3.30	3.71	0.72	7.16

Ion	K^+	La^{3+}	Li^+	Mg^{2+}	Na^+	O^{2-}	Rb^+	Sr^+	Ti^{4+}
α	1.48	1.16	0.03	0.10	0.16	4.31	2.20	1.78	0.21

From L. Pauling, *Proc. Roy. Soc.*, **A114**, 181 (1927) and Tessman, Kahn and Shockley, *Phys. Rev.*, **92**, 890 (1953).

Anions, being large, are more polarizable than most cations, and small cations hardly contribute to the polarizability at all. Adjustment of the refractive index of a material, such as an optical fibre, is made possible by incorporating ions of greater or lesser polarizability into the structure, for example GeO_2 can be added to SiO_2 to increase the refractive index of a quartz (silica) fibre. The Ge^{4+} ion is 29 per cent larger than the Si^{4+} ion, so according to equation (8.2) we might expect its polarizability to be a bit more than double. Very fine adjustments in refractive index are possible by control of composition, so that both stepped-index and graded-index optical fibres may be manufactured (see chapter 9).

8.6 Energy Absorption in Dielectrics

At low frequencies, in the radio-frequency region of the electromagnetic spectrum, the dipolar molecules in some media may be capable of rotation in response to the oscillations of an alternating electric field. This rotation will be opposed by damping forces, such as viscous drag, which results in a characteristic relaxation time for the dipoles, which is given by

$$\tau = 1/\omega_0$$

where ω_0 is the frequency of maximum power absorption, the resonant frequency. When $\omega \ll \omega_0$, the dipoles are in phase with the field and no energy is absorbed and when $\omega \gg \omega_0$, the dipoles cannot respond to the field at all and again no energy is absorbed. At low frequencies the dipoles make maximum contribution to the permittivity and at high frequencies they make no contribution.

8.6.1 Dielectric Relaxation: The Debye Equation

We can write the time dependence of the polarization as

$$P(t) = P_0\exp(-t/\tau)$$

where τ is the relaxation time. To obtain the corresponding function, $P(\omega)$, in the frequency domain, we perform a Fourier transform in the positive frequency domain only:

$$P(\omega) = \int_0^\infty P_0\exp(-t/\tau)\exp(-j\omega t)dt$$

$$= P_0/(\omega_0 + j\omega)$$

where we have substituted ω_0, the resonant frequency of the relaxation process, for $1/\tau$. Now P depends directly on the permittivity, so instead of a complex polarization we shall use a complex permittivity of the same form:

$$\epsilon(\omega) = A/(\omega_0 + j\omega) + B \tag{8.7}$$

A and B are constants which can be found by considering what happens at $\omega = 0$ and $\omega = \infty$. The permittivity at zero frequency is the static permittivity, ϵ_s, so putting $\omega = 0$ in (8.7) leads to

$$\epsilon(0) = \epsilon_s = A/\omega_0 + B$$

At frequencies far above resonance, the permittivity is ϵ_∞, which is less than the static permittivity. Substituting $\omega = \infty$ in (8.7) gives $B \equiv \epsilon_\infty$, so that $A \equiv \omega_0(\epsilon_s - \epsilon_\infty)$ and (8.7) becomes

$$\epsilon(\omega) = \frac{\epsilon_s - \epsilon_\infty}{1 + j(\omega/\omega_0)} + \epsilon_\infty \tag{8.8}$$

The complex permittivity can be expressed as

$$\epsilon = \epsilon' - j\epsilon''$$

which can be compared to (8.8) yielding

$$\epsilon' = \epsilon_\infty + \frac{\epsilon_s - \epsilon_\infty}{1 + (\omega/\omega_0)^2}$$

and

$$\epsilon'' = \frac{(\omega/\omega_0)(\epsilon_s - \epsilon_\infty)}{1 + (\omega/\omega_0)^2}$$

The imaginary part of the permittivity, ϵ'', causes energy to be absorbed and is a maximum at the resonant frequency, ω_0, and zero far above and below it. The real part of the permittivity starts off at low frequencies at the static value, falls as the resonant frequency is approached to reach halfway between ϵ_s and ϵ_∞ at $\omega = \omega_0$, then drops to ϵ_∞ as the frequency increases further. Figure 8.6 shows the graphs of ϵ' and ϵ'' against ω.

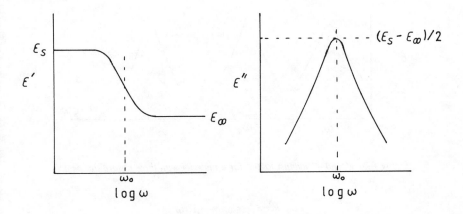

Figure 8.6 ϵ' and ϵ'' against $log(\omega)$

If there are several absorption mechanisms in the material, such as orientational, molecular and electronic relaxation processes, then there are corresponding peaks in ϵ'' and reductions in ϵ' as the frequency goes up. As electrons can respond very quickly to electric field changes, the absorption peak for electronic polarization occurs at very high frequencies – in the ultra-violet or optical region. Ionic motion is several thousand times slower and so ionic polarization leads to peaks in absorption in the infra-red region, while orientational polarization gives rise to peaks at UHF and microwave frequencies. Figure 8.7 shows a plot of ϵ'' and ϵ' against ω for a hypothetical solid exhibiting all three types of absorption.

8.6.2   *The Loss Tangent*

The *loss tangent*, $\tan\delta$, is equal to ϵ''/ϵ'. In chapter 2, Maxwell's equations were used to find the energy absorbed from electromagnetic waves by conducting media and a dispersion relation was derived – equation (2.36):

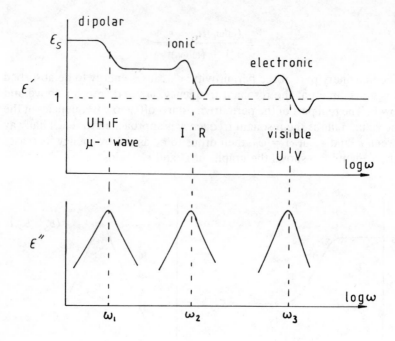

Figure 8.7   *ε′ and ε″ against log(ω) for a material with three absorption peaks*

$$k^2 = \mu\epsilon\omega^2 + j\mu\sigma\omega$$

In this equation, ϵ is real ($= \epsilon_0\epsilon_r$). The phase of k^2 is identical to δ (see problem 2.8), so that

$$\tan\delta = \sigma/\epsilon\omega = \epsilon''/\epsilon'$$

Since this ϵ is real, we can identify it with ϵ' and then $\epsilon'' \equiv \sigma/\omega$. Now power losses per unit volume in a conducting medium are given by

$$W_v = \sigma\mathscr{E}^2$$

Now if $\epsilon'' \equiv \sigma/\omega$, then

$$W_v = \epsilon''\omega\mathscr{E}^2$$

Now $\epsilon'' = \epsilon'\tan\delta$ so the power losses/m³ are

$$W_v = \omega\mathscr{E}^2\epsilon'\tan\delta \tag{8.9}$$

A word of caution here might be advisable on the nomenclature, which is confusing. The *dissipation factor* is the same as $\tan\delta$, the *loss tangent*. The *loss factor* is ϵ'' and is therefore equal to $\epsilon'\tan\delta$. The *power factor* is $\sin\delta \approx \tan\delta$, a very good approximation for capacitor materials.

Because dielectric losses are proportional to $\mathscr{E}^2$, they are an important consideration in the design of capacitors and high-voltage cables. Though the frequency of the National Grid is only 50 Hz, the fields in the dielectric of a 300 kV cable may be 15 MV/m, so if tanδ is 0.01, the losses will be

$$W_v = 2.3 \times 8.85 \times 10^{-12} \times (15 \times 10^6)^2 \times 2\pi \times 50 \times 0.01 \approx 14 \text{ kW/m}^3$$

The loss is significant, though much less than the conductor losses. In practice, solid dielectrics cannot be used above about 33 kV because of the large thickness required and consequent problems in dissipating heat. The field in solid dielectrics for power cables is limited to about 5 MV/m to lessen the risk of breakdown.

Dielectric heating relies on tanδ losses to heat plastics or to dry wood. The frequencies used are restricted by law in Britain to 13.56 MHz, 27.12 MHz, 896 MHz and 2.45 GHz. One advantage of dielectric heating is that the heat is produced uniformly throughout the material, rather than just within the skin depth as in RF-heated, metallic bodies.

8.6.3 An Example of Capacitor Losses

Table 8.2 gives some relative permittivities and loss tangents for a variety of dielectrics used in making capacitors. The properties of dielectrics, especially tanδ, are very dependent on the preparation of the material, as

Table 8.2   Properties of Dielectrics at 300 K and 1 MHz

Material	ϵ_r	$tan\delta \times 10^{-4}$
Alumina[a]	10	5–20
Porcelain[b]	5	75
Silica (Quartz)	3.8	2
Mica[c]	5.4	3
$BaTiO_3$	500	150
PZT4[d]	1000	40
Nylon 610	3.1	220
Perspex	2.6	145
Polycarbonate[e]	3.1	10
Polyethylene	2.3	2
Polystyrene	2.6	0.7
PTFE	2.1	2
PVC	3	160

[a] Low porosity, polycrystalline.
[b] High voltage, insulator grade.
[c] Muscovite.
[d] Lead zirconate–titanate, $Pb(Zr,Ti)O_3$.
[e] Capacitor grade.

well as the temperature (tanδ tends to increase with temperature rise) and frequency, so the values given in the table should be treated as guidelines only.

Suppose we want to make a capacitor with a capacitance, C, of 1 μF working at an rms voltage of 50 V at 1 MHz with a dielectric thickness, d, of 50 μm. The electric field is therefore 1 MV/m and if we choose polycarbonate as the dielectric its area can be found from

$$C = \epsilon A/d$$

or

$$A = Cd/\epsilon$$

Substituting for $C = 1$ μF, $d = 50$ μm and $\epsilon = 3.1\epsilon_0$ leads to $A = 1.8$ m^2 so that the volume of the polyethylene (Ad) is 9×10^{-5} m^3 (90 cc). From equation (8.9) the losses will be 170 kW/m^3, or 15 W. Repeating the calculation using PZT4 leads to a volume of 2.8×10^{-7} m^3 (0.28 cc) and a power dissipation of 62 W. The power dissipated is proportional only to tanδ, as the product of capacitor volume and power loss per unit volume is independent of ϵ. PZT4 is therefore useless for high frequency capacitors, while polycarbonate will just do at 1 MHz. Polystyrene would be a better choice.

8.7 Dielectric Breakdown

The breakdown of dielectrics in high fields is bedevilled by particular circumstances rather than being illuminated by general scientific principles – though these have their place – because breakdown often accompanies a flaw in the material rather than being dependent on its inherent properties. Three types of dielectric breakdown are distinguishable:

- Intrinsic
- Thermal
- Discharge

Intrinsic breakdown is caused by the acceleration of free electrons in high fields which can ionize other atoms to cause an avalanche effect. The original electron must acquire roughly the band-gap energy in order to promote electrons from valence band to conduction band. Now the electron has to gain energy between collisions, so that if its mean free path is l_m, this energy is $\mathscr{E}el_m$, while the energy required is eE_g, where E_g is the band gap in eV. Equating the two energies gives $\mathscr{E} = E_g/l_m$. Now E_g will be about 5 eV, while $l_m \approx 50$ nm, giving the intrinsic breakdown field as 100 MV/m. In some cases, such as amorphous silicon nitride films, where

l_m is small, this figure can be exceeded by a factor of five, giving intrinsic breakdown fields as high as 500 MV/m, or 500 V/μm. Breakdown fields of this magnitude cannot be achieved in practical components because of imperfections in the dielectric.

Thermal breakdown occurs when tanδ losses cause heating which lowers the breakdown field. Each dielectric will have a temperature limit which cannot be exceeded without risking thermal breakdown.

Discharge breakdown occurs when the gas in small pockets in the dielectric becomes ionized by the field. The gaseous ions are accelerated by the field and impact the side of the cavity causing damage and more ionization. Because gases are more readily ionized than solids, voids in dielectrics must be avoided at all costs. In high-voltage power cables the problem can be reduced by permeating the insulation with a difficult-to-ionize, high-pressure gas.

8.8 Ferroelectrics

A ferroelectric material possesses a spontaneous polarization which can be aligned by an external field. Hysteresis occurs so that a P–$\mathscr{E}$ loop may be drawn for a ferroelectric, as in figure 8.8(a), exactly like the J–H loop of a ferromagnet. The spontaneous polarization-temperature graph is like a ferromagnet's (figure 8.8(b)) and the spontaneous polarization vanishes at the Curie temperature also, above which the material is said to be

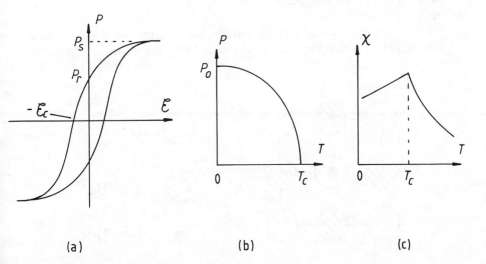

Figure 8.8 *(a) Polarization against $\mathscr{E}$ for a ferroelectric; (b) spontaneous polarization against T for a ferroelectric; (c) electric susceptibility against T for a ferroelectric*

paraelectric (see figure 8.8(c)). There are two classes of ferroelectric: the order–disorder type involving the movement of hydrogen atoms, for example KH_2PO_4, and the displacive type such as the perovskites, typified by $BaTiO_3$. Only the latter will be discussed as they are by far the most important. (These materials or derivatives from them have recently been found to be superconductive at extraordinarily high temperatures – see chapter 10.)

Perovskite is a mineral with the formula $CaTiO_3$, whose structure is shown in figure 8.9. Essentially the structure consists of a close-packed anion lattice in which every fourth anion is replaced by a divalent cation (the face centres in figure 8.9), with a quadravalent cation in the octahedral site at the centre of the unit cell. The first important perovskite ferroelectric was $BaTiO_3$, but its relatively low Curie temperature (≈ 400 K) led to the development of superior materials, especially $Pb(Zr, Ti)O_3$, which has $T_c \approx 650$ K. The magnitude of the ionic displacements in the unit cell can be calculated from the saturation polarization, P_s (which is the dipole moment/m^3). The unit cell of $BaTiO_3$ is cubic above the Curie temperature and the lattice parameter is about 4 Å. At 300 K the saturation polarization is 0.26 C/m^2, so the dipole moment per unit cell, μ, is $0.2 \times (4 \times 10^{-10})^3$, which is 1.7×10^{-29} C m. The charge on the Ti^{4+} and Ba^{2+} ions is $6e$ (= 9.6×10^{-19} C) per unit cell, so their displacement with respect to the O^{2-} ions, ξ (= μ/q), must be $1.7 \times 10^{-29}/9.6 \times 10^{-19}$, or 1.8×10^{-11} m (0.18 Å). This displacement causes the structure to become tetragonal below T_c, though the axial ratio is close to unity.

8.8.1 The Catastrophe Theory

Displacive ferroelectrics can be interpreted in terms of a polarization catastrophe in which the local field produced by the displacement of the

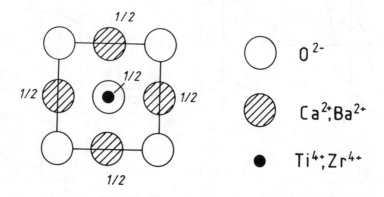

Figure 8.9   *The perovskite unit cell*

ions is larger than the restoring forces, so that the displacement, and hence the polarization, tends to infinity. The non-linearity of the restoring forces results in a finite, though very large polarization. While deriving the Clausius–Mossotti relation, equation (8.6), we found equation (8.5):

$$\epsilon_r = (1 + 2\beta)/(1 - \beta)$$

where $\beta \equiv \Sigma \; \alpha N_v/3\epsilon_0$, the sum of the contributions of all the ions in the structure to the polarization. Thus when $\beta = 1$, the relative permittivity becomes infinite. Since the derivation of equation (8.5) relies on an isotropic, or cubic, ionic environment, the structural distortion as β approaches unity restricts the value of ϵ_r. Nevertheless, the relative permittivity reaches extraordinarily high values (over 10 000) in some perovskites.

It may be shown (see problem 8.4) that in the paraelectric state $(T > T_c)$, the relative permittivity is given by

$$\epsilon_r = 1/\zeta(T - T_c) \tag{8.10}$$

where ζ is the coefficient of linear thermal expansion. Thus the susceptibility suffers a rapid fall just above the Curie point as in figure 8.8(c).

8.8.2 Uses of Ferroelectrics

Ferroelectrics, somewhat surprisingly, find wide use as sound transducers (1 kHz to 10 MHz, for both sonar and ultrasonic applications), accelerometers, delay lines, positional transducers, strain gauges, in spark ignition (domestic gas appliances) and pressure gauges, besides the more obvious application as a capacitor material. Most of these applications are the consequence of their large electrostrictive coefficients.

When an electric field is applied to an electrostrictive material it contracts or expands, depending on the sign of the coefficient, along the polarization direction. The polarity of the field is immaterial, unlike a piezoelectric material such as quartz (once used in sound transducers, until superseded by magnetostrictive nickel transducers and then by ferroelectrics). Thus, if an alternating field is applied to an electrostrictor, it responds as in figure 8.10(a): there is frequency-doubling, which can be removed by applying a bias field as in figure 8.10(b). (The same must be done with magnetostrictive sound transducers, biasing being accomplished conveniently with a permanent magnet.) The advantage of ferroelectric materials lies in the possibility of their having a built-in biasing field, which can be produced by a process called *poling*. The ceramic is heated to a little below T_c and a large field (perhaps 1–5 MV/m) is applied for a few minutes. The ferroelectric then has a remanent polarization which can only be destroyed by heating to near T_c or employing fields in excess of the coercive field (about 1 MV/m). After poling, the ferroelectric behaves just

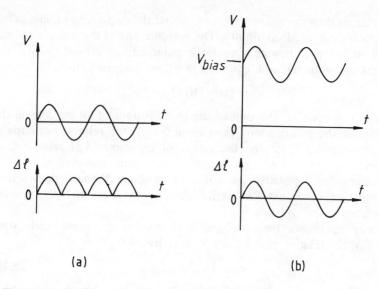

Figure 8.10   *(a) Frequency-doubling in an electrostrictive material; (b) the response of an electrostrictor to an applied field when a large bias field is also present*

like a piezoelectric material, though strictly it is a polarized, electrostrictive material.

Ferroelectric materials are also good pyroelectric materials. The pyroelectric coefficient is defined by

$$\lambda = dP/dT$$

A change in temperature of the material causes a change in polarization and hence a change in surface charge. In a 'good' material, such as barium niobate, $\lambda \approx 3 \times 10^{-3}$ C/m²/K, and as a charge of 10^{-16} C is detectable, it can be seen that pyroelectric devices can register temperature changes of the order of 1 μK. Extensive use is made of this in military hardware responding to infra-red radiation (heat). Thermal-imaging devices are also made from pyroelectric materials

Problems

8.1. A certain material has a single maximum in the imaginary part of its relative permittivity (ϵ''_r) at 1 GHz. Its static relative permittivity is 6 and its refractive index in the infra-red region is 1.732. What is the maximum value of tanδ and at what frequency is this maximum?
[*Ans. 0.354, 1.41 GHz*]

8.2. α-quartz (SiO_2) has a density of 2650 kg/m³. Estimate its refractive index using data from table 8.1. Why does this value not agree with the measured value of 1.57 at 590 nm?
[Ans. 4.1]

8.3. At room temperature the static relative permittivity of water is 80. A plot of tanδ against frequency shows a maximum at 30 GHz of 3.3. Deduce the refractive index of water in the visible spectral region and the relaxation time for water dipoles. How would you expect the relaxation time, frequency of maximum tanδ and refractive index to vary from − 1°C to 20°C?
[Ans. 1.33, 3.6 × 10⁻¹¹ s]

8.4. Above its Curie point, strontium titanate undergoes thermal expansion leading to a reduction in ionic concentration (N_v). If the polarizability of the ions in $SrTiO_3$ is constant, show that the relative permittivity above T_c is given by equation (8.10). Using the data for $SrTiO_3$ given below, find ζ and T_c.

$T(K)$	140	160	180	200	220	240
ϵ_r	700	580	500	435	390	350

[Use $V = V_0 (1 + 3\zeta\Delta T)$, where V is the volume, V_0 the original volume, ΔT is the change in temperature and ζ is the coefficient of linear thermal expansion, to find the change in N_v, and thus β, with T, then substitute for β in equation (8.5)]
[Ans. 1.4 × 10⁻⁵/K, 40 K]

8.5. Show that the energy stored in a capacitor per unit volume is given by

$$U = \tfrac{1}{2}\epsilon_r\epsilon_0\mathscr{E}^2$$

where $\mathscr{E}$ is the applied field, V/d, where V is the voltage on the capacitor plates and d is the thickness of the dielectric. If the maximum permissible field is 5 MV/m, what volume of capacitor with $\epsilon_r = 1000$ is required to store 1 kWh? Compare this to the energy storage capabilities per unit volume of an inductor, a lead–acid battery and petrol. How much water must fall through 500 m (as at the Dinorwic pumped storage scheme) to produce 1 kWh of electricity?
[Ans. 33 m³; 1 kWh/m³, 80 kWh/m³, 10 MWh/m³; 720 kg]

8.6. Suggest how half the Ti^{4+} ions in $BaTiO_3$ may be replaced by Sc^{3+} ions, without excessive distortion. Is there more than one way? If so, say what they are. Some ionic radii are given below

Ion	O^{2-}	S^{2-}	Se^{2-}	F^-	Cl^-	Br^-	K^+	Rb^+	Pb^{2+}
Radius (Å)	1.40	1.84	1.98	1.36	1.81	1.96	1.33	1.49	1.20

Ion	Sr^{2+}	Ba^{2+}	Sc^{3+}	Y^{3+}	La^{3+}	Fe^{3+}	Ti^{4+}	Zr^{4+}	Ce^{4+}
Radius (Å)	1.12	1.36	0.73	0.89	1.06	0.64	0.61	0.72	0.90

8.7. The following data relate to liquid $CHCl_3$ (chloroform). Do they fit the Langevin function? Calculate the dipole moment.

ϵ_r	6.76	6.12	5.61	4.81	3.71	3.33	2.93
$T(°C)$	-60	-40	-20	20	100	140	180

[Ans. 2.6 D]

8.9. The dipole moment of the KCl molecule in the gaseous state is 10.3 D. If the ionic radii are as given in problem 8.6, what will be the charge on the ions as a fraction of the electronic charge? What would you estimate for the dipole moment of RbBr? What will be the dielectric constant of KCl gas at 2000 K and one atmosphere pressure?

[Ans. 0.69e; 11.3 D; 1.0059]

9

Materials for Optoelectronics

In this chapter the term optoelectronics will embrace devices which convert electrical signals into optical ones and vice versa, as well as the medium of transmission. Though it can be argued convincingly that optoelectronics has existed in practice for some considerable time, it was the remarkable advances in materials for optical fibres – especially for telecommunications – occurring in the 1970s that led to the current wealth of activity in this field. Of course, significant developments in solid-state devices for the production and detection of light would have taken place without the advent of the near-lossless fibre, but not to the same degree.

For the whole of this chapter, λ refers to the wavelength of light in a vacuum *only*. In any other medium the wavelength will always be given as λ/n, where n is the refractive index (RI). Visible light has a (vacuum) wavelength from roughly 800 nm to 400 nm, or about an octave in frequency. Initially, the wavelength used in fibre-optic communications was that of a GaAs LED working at 0.84 μm, but since silica (quartz) fibres have a minimum in their losses at about 1.3 μm, this became the preferred wavelength in the early 1980s, though the light is not in the visible range. Subsequent developments have led to the use of a wavelength of 1.55 μm, and soon further material improvements (such as the use of fibres made of fluorides which are more transparent to infra-red) look like increasing this still further. Here the optical range of wavelengths is taken to be from 3 μm to 200 nm, roughly a decade in frequency, and including the near infra-red and near ultra-violet.

9.1 Light-emitting Diodes (LEDs)

LEDs play two distinct roles in optoelectronics: as display devices and for signal generation and modulation in communications.

9.1.1 LEDs for Displays

LEDs operate by the emission of photons produced by the transition of electrons from conduction to valence band. This transition occurs in the material near a p–n junction which is supplied with an excess of minority carriers by a forward-biasing current. The material for a visible LED must have a bandgap of at least 1.7 eV so that the transition causes the emission of visible light. A glance at table 4.1 shows that III–V compounds such as GaP and II–VI compounds such as CdS are suitable materials, and in fact the majority of LEDs for displays are made from III–V compounds of the general formula $GaAs_{1-y}P_y$ or $Ga_{1-x}Al_xAs$, which can cover the bandgap range from 1.42 to 2.24 eV (880–558 nm), that is, red to yellow-green. Some devices made from SiC have been made which emit blue light, but at low efficiency. II–VI compounds do not find favour as they are hard to dope p-type. However, as can be seen from table 4.1 too, while GaAs has a direct gap, both AlAs and GaP have indirect gaps. Because the emission of a photon in an indirect-gap material also requires a phonon to participate during the electron's interband transition, the photon probability (or quantum efficiency) is low and the light intensity is low too. This fall in efficiency is compensated to some extent by the increase in the human eye's response at wavelengths around 555 nm (the wavelength of maximum response, with a bandwidth of about 100 nm, see problem 9.1). Doping the material with nitrogen introduces an electron-trapping level about 10 meV below the conduction band, which enhances the efficiency, as shown in figure 9.1.

The direct bandgap in GaP is about 2.74 eV, while the indirect bandgap in GaAs is about 1.81 eV and we see from figure 9.2 that the borderline composition between direct and indirect gaps is at the point where direct and indirect gaps are equal. Assuming a linear relationship between composition and bandgap leads to

$$E_g(\text{direct}) = 1.32y + 1.42 \qquad (9.1a)$$

$$E_g(\text{indirect}) = 0.44y + 1.81 \qquad (9.1b)$$

E_g is in eV. Solving for equal gap energies gives $y = 0.44$ at $E_g \approx 2$ eV (625 nm, orange light). Without nitrogen doping, the quantum efficiency of the material falls rapidly as the mole fraction of phosphorus approaches 0.44, so $y = 0.4$ was a popular choice, leading to emission at about 630 nm for a red LED (see also problem 9.4).

The light emitted from a LED is not monochromatic since the electrons in the conduction and valence bands have a distribution of energies (given by $F(E)D(E)$, see chapter 3), which results in maximum output at a wavelength in μm given by

$$\lambda = 1.24/(E_g + k_BT) \qquad (9.2)$$

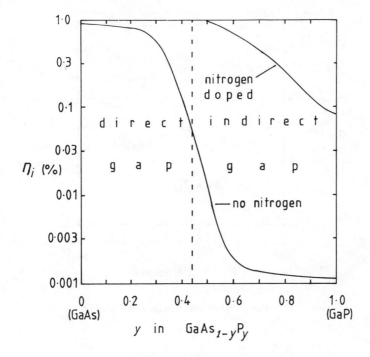

Figure 9.1 *The effect of nitrogen on* η_j *in Ga(As,P)*

with E_g and $k_B T$ in eV. The spread of energies for the emitted photons is about 3 $k_B T$ at the − 3 dB points of the spectrum. Taking GaAs$_{0.6}$P$_{0.4}$ as an example, its gap energy is 1.95 eV from equation (9.1a), so from equation (9.2) its maximum output is at 628 nm at 300 K. Taking $1\tfrac{1}{2}k_B T$ either side of 1.976 eV gives the half power wavelengths as 640 and 615 nm, a spectral width of 25 nm. The spectral width goes as the square of the wavelength of maximum emission, so infra-red LEDs have rather broad spectral responses (see problem 9.2).

The structure of a LED chip is shown in figure 9.3. It consists of a p–n junction formed by diffusion of zinc (p-type dopant). If the material is GaAs$_{0.6}$P$_{0.4}$, then the substrate is GaAs or GaP onto which a graded layer of GaAs$_{1-y}$P$_y$ is grown by CVD, using GaCl from Ga/HCl to transport the gallium and PH$_3$ and AsH$_3$ to transport the phosphorus and arsenic. The graded layer starts at $y = 0$ (for a GaAs substrate) or $y = 1$ (for a GaP substrate) and ends at $y = 0.6$ to minimize lattice mismatch at the interface. An n-type GaAs$_{0.6}$P$_{0.4}$ layer (about 1 μm thick) is then grown on top of the graded layer using selenium or tellurium as a dopant. The upper contact to the p-type region is made so that as much light is emitted as possible from the top surface.

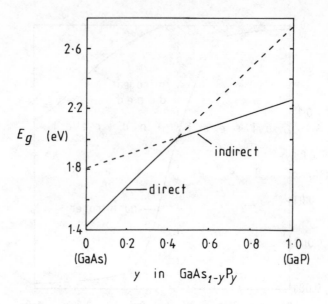

Figure 9.2   *Direct and indirect bandgaps in Ga(As,P)*

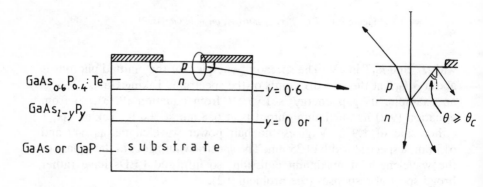

Figure 9.3   *An LED chip*

When the substrate is GaAs, which is opaque (having a smaller bandgap than the material of the p–n junction), photons emitted down from the junction, or photons reflected from the top interfaces, are absorbed. A GaP substrate is transparent (having a larger bandgap than the junction material) to these photons, so by making the bottom contact reflective they can be returned to the sides or top.

The Efficiency of LEDs

The efficiency of an LED can be defined by the equation

$$\eta = \text{Optical power out/electrical power in}$$

which is approximately the same as

$$\eta = \text{No. of photons emitted/No. of recombinations}$$

η can be divided into two components: η_i, the internal, or quantum, efficiency; and η_e, the external efficiency, so that $\eta = \eta_i \eta_e$. The quantum efficiency depends on the type of transition (direct or indirect) and whether there are intermediate levels in the bandgap; and even though a direct transition may be achieved, it may not produce an optical photon, with the result that quantum efficiencies are small – less than one per cent (see figure 9.1).

The external efficiency arises because there is a problem in getting the light out of the device. Part of the problem is caused by the refractive index (RI) mismatch between air and the LED material. It can be shown that the fraction of light energy transmitted at a boundary between two media is given by

$$T = 1 - R = 1 - \left[\frac{n_1\cos\vartheta_1 - n_2\cos\vartheta_2}{n_1\cos\vartheta_1 + n_2\cos\vartheta_2} \right]^2 \tag{9.3}$$

where n_1, n_2 are the RIs of the two media and ϑ_1, and ϑ_2 are the angle of incidence and the angle of refraction as shown in figure 9.4. T is known as the *transmissivity* and R is the *reflectivity*, ϑ_1 and ϑ_2 are related by Snell's law:

$$n_1\sin\vartheta_1 = n_2\sin\vartheta_2$$

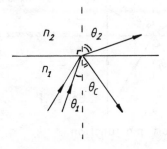

Figure 9.4 *Refraction and reflection at an interface*

The RI of a Ga(As,P) LED is about 3.4 (= n_1), so at normal incidence ($\vartheta_1 = 0$) with air as the second medium ($n_2 = 1$), $T = 0.7$. But matters are made much worse by the existence of a critical angle of incidence (ϑ_c), above which all the incident light is reflected. The critical angle is found by putting $\vartheta_2 = 90°$ in Snell's law, so that $\vartheta_c = \sin^{-1}(n_2/n_1)$, which in this case is 17°, a fairly small angle. Therefore, unless the light emitted from the diode lies within a cone of a semi-angle, ϑ_c, it cannot emerge from the diode. A cone of small semi-angle, ϑ_c, encloses a solid angle of about $\pi\vartheta^2_c$, so the fraction of light emerging from the diode is $T \times \pi\vartheta^2_c/4\pi$, or $\tfrac{1}{4}\vartheta^2_c T$. Now $\vartheta_c \approx n_2/n_1$, for small ϑ_c, and the fraction of light emitted, assuming all reflected light is absorbed, is $\tfrac{1}{4}(n_2/n_1)^2 T$. Thus η_e is only about $1\tfrac{1}{2}$ per cent for an unencapsulated Ga(As, P) LED. By encapsulating the LED chip in transparent epoxy plastic with a RI near to the square root of the RI of the LED material (see problem 9.3), the reflection losses are reduced. Rounding the outer surface of the epoxy helps to reduce angles of incidence to less than ϑ_c (see figure 9.5).

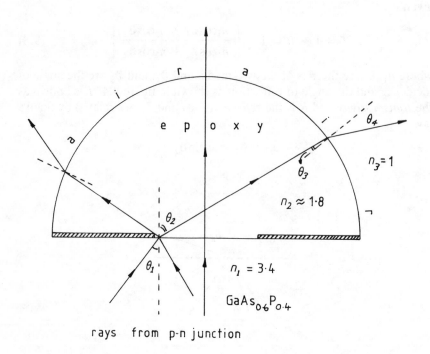

rays from p-n junction

Figure 9.5 *Refraction with a curved matching layer of epoxy*

9.1.2 Signal LEDS

Signal LEDs must be capable of rapid modulation at a wavelength which suffers least absorption by the medium of transmission, about 1.3 μm in the case of silica fibres. Binary pulse-code modulation (PCM) is the usual modulating technique and its bandwidth is determined by the recombination lifetime, τ_r, of the minority carriers in the LED. τ_r is reduced by heavy doping to about 10–20 ns, giving data rates of the order of $1/2\tau_r$, or 25–50 Mbits/s. Looking at the gap wavelengths, λ_g, of table 4.1 we see that silicon would be suitable, but it has an indirect gap and is very inefficient. Ga(As, Sb) or In(P, As) would be good choices as they would have direct gaps and presumably be more efficient. In practice a quaternary composition, $In_{1-x}Ga_xAs_{1-y}P_y$, is used with $x = 0.2$ and $y = 0.7$ in order that there can be a good lattice match to an InP substrate. Figure 9.6 shows the relation between bandgap and lattice parameter for some III–V compounds, from which appropriate bandgaps and lattice parameters can be chosen. Hence the origin of the term *bandgap engineering*. (See also problem 9.4.)

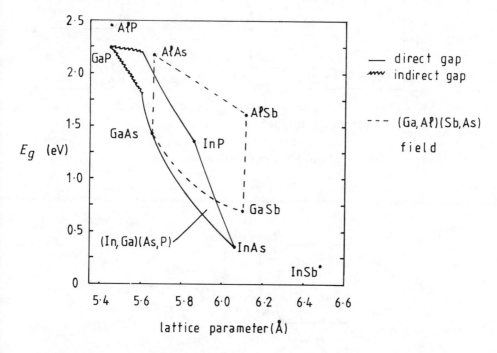

Figure 9.6 E_g *against lattice parameter for III–V compounds*

Heterojunctions

If a p–n junction is made from different materials, such as p-type (In, Ga)As and n-type InP, it is said to be a heterojunction while a p–n junction in silicon could be termed a homojunction. The important point about heterojunctions is the differing bandgaps of the constituents. If the p-type material has a larger bandgap than the n-type material, then a P-n junction is formed: upper case letters are used for the material with the larger bandgap. Figure 9.7 shows the energy-level diagram for a P–n junction, from which it can be seen that the energy barrier is greater than it would be in a p–n junction. By using a double heterojunction structure such as N–n–P, as in figure 9.8, it is possible to confine photon production to the central n-type region. The concentration of minority carriers in this region is larger than it would be in a P–n junction because of the increased

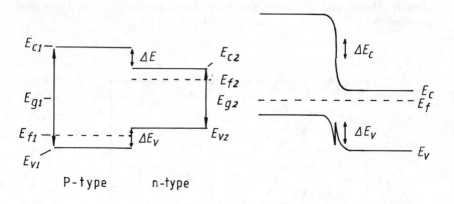

Figure 9.7   *Energy levels at a heterojunction*

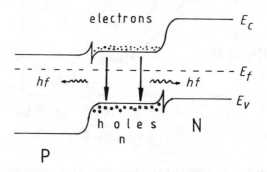

Figure 9.8   *Energy levels at a double heterojunction*

barriers to hole movement in the valence band and electron movement in the conduction band. Thus the light intensity is enhanced and its wavelength is determined by the smaller bandgap. Solid-state lasers are also made with double heterojunctions.

Two type of signal LED are made: the surface-emitting LED (or Burrus LED, after its inventor) and the edge-emitting LED. The Burrus LED is shown in figure 9.9. The epoxy holding the LED onto the end of the quartz fibre is index-matched to reduce interfacial losses (see problem 9.3) and so provide good optical coupling.

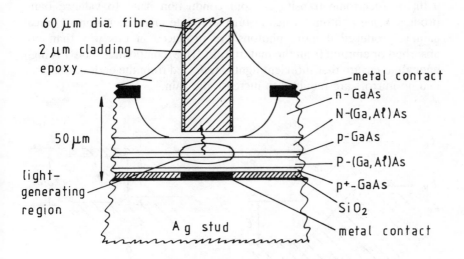

Figure 9.9 *The Burrus LED*

9.2 Solid-state Lasers

Laser is an acronym: Light Amplification by the Stimulated Emission of Radiation. In a LED, light is emitted spontaneously as current flows; in a p–n junction laser, light is emitted during the photon-stimulated transition of electrons across the bandgap. Solid-state lasers can be made very small so that efficient coupling into small optical fibres is possible. They can operate continuously at room temperature for years. Besides being used as signal sources in optical communications they have found use in laser printers, in compact-disc players and elsewhere.

Lasing occurs when electrons occupying a higher, non-equilibrium, energy state fall into a lower energy state (E_1). The transition is accompanied by the emission of a photon of wavelength, λ, given by

$hc/\lambda = (E_2 - E_1)$. An essential requirement for lasing is the existence of electrons in a higher energy state, called a population inversion. If electrons in large numbers can be injected into p-type material where there are large numbers of holes available for recombination, then a population inversion exists.

Heavily doping a p^+/n^+ junction on both sides forces the fermi level into both conduction and valence bands as in figure 9.10(a): the material is said to be degenerate. Then, under high forward bias ($\approx E_g$, volts) electrons are injected across the junction creating a population inversion in a very small region (the active region) near the junction, as in figure 9.10(b). Electronic transitions from conduction band to valence band produce some photons, which in turn stimulate more transitions. Optical gain is produced if more photons are produced in this way than are absorbed or emitted from the material. In III–V compounds the RI is high enough to ensure that sufficient light is reflected from the ends of the laser and no mirrors are required to increase the gain.

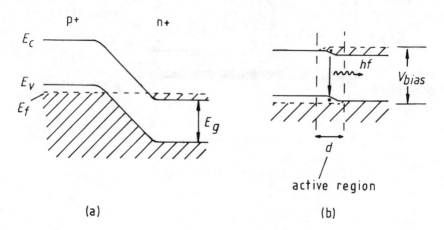

Figure 9.10   *(a) Energy levels at an unbiased p^+-n^+ junction; (b) energy levels at a forward-biased p^+-n^+ junction*

The light output is confined to the active region in a homojunction laser, whose width is approximately the diffusion length of electrons in the p-type material, about 2 μm (see figure 9.11(a)). Because the confinement is relatively poor, gain is reduced and large current densities are required before lasing begins. These large current densities mean that much power is dissipated and the laser cannot be used continuously or it overheats. Using a double heterojunction (for example, $P^+-P-p-N$) for lasing has three advantages: first, the RI either side of the active region is reduced (higher bandgap means lower RI) so that an optical waveguide is obtained;

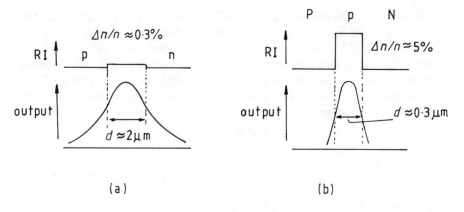

Figure 9.11   *(a) RI and light output at a p–n junction; (b) RI and light output at a*
P–p–N double heterojunction

second, threshold current densities are lowered, making cw operation possible; and third, the active region can be made small by reducing the thickness of low-bandgap material, for example to 0.3 μm (see figure 9.11(b)). All these improvements result in much lower power consumption (as little as 50 mW). Figure 9.12 shows a buried, double heterojunction laser in cross-section. Replacing some of the Ga in GaAs with Al, as in $Ga_{0.7}Al_{0.3}As$, increases the bandgap (see figure 9.6) and reduces the RI by about 5 per cent. The central GaAs region in figure 9.12 thus forms a rectangular optical waveguide.

Two important parameters for a laser are its total attenuation coefficient, α_t, and its optical gain, g. α_t is made up of two parts: the part due to scattering in the material, α_s, and the end losses due to reflection from the laser crystal/air interface. Consider a laser of length, L, in which light of

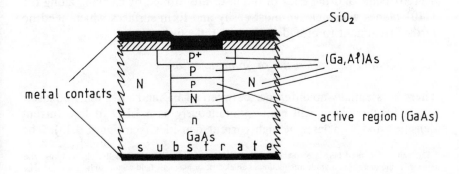

Figure 9.12   *A buried, double heterojunction laser*

intensity, I_0, travels from one end to the other, where it is reflected. The intensity of the light when it reaches the other end will be $I_0\exp(-\alpha_s L)$. If the reflectivity of this end of the laser is R_1, then the reflected light will have an intensity of $I_0 R_1\exp(-\alpha_s L)$. On returning to the end it started from, the light is subject to further attenuation and its intensity will be $I_0 R_1\exp(-2\alpha_s L)$. If the reflectivity at this end is is R_2, then the intensity for the complete round trip will be

$$I = I_0 R_1 R_2 \exp(-2\alpha_s L)$$

Taking logs:

$$\ln(I/I_0) = -2\alpha_s L - \ln(1/R_1 R_2)$$

Expressing this per unit length of laser:

$$(1/2L)\ln(I/I_0) = -\alpha_s - (1/2L)\ln(1/R_1 R_2)$$

So the attenuation per unit length for a round trip becomes[*]

$$\alpha_t = \alpha_s + (1/2L)\ln(1/R_1 R_2) \tag{9.4}$$

Now as photons move through the laser, they are not only being annihilated by α_t, they are also stimulating emission of more photons, producing a laser gain, g. For the laser to work, $g \geq \alpha_t$.

For instance, suppose $g = 2500 \text{ m}^{-1}$, while $\alpha_s = 600 \text{ m}^{-1}$. The laser material is of RI 3.5. What is the *minimum* length for lasing to occur?

The transmissivity, T, for normal incidence at the ends of the laser, is found from equation (9.3) as 0.69, so R $(= R_1 = R_2)$ is 0.31. Putting $\alpha_t = 2500$ and $\alpha_s = 600$ into (9.4) yields

$$1900 = (1/2L)\ln(10.4)$$

leading to $L = 616$ μm as the minimum length for the laser, a typical value for such devices. This figure could be reduced by silvering the ends of the laser, which would also reduce the output. The junction plane is usually a {100} plane and the ends of the laser are formed by cleaving along the {110} planes which cleave most easily and form surfaces which need no further treatment to provide sufficient reflectivity.

9.2.1 Output Characteristics of Solid-state Lasers

There is essentially no difference between a LED and a p–n junction laser, and a junction laser will emit spontaneously as a LED at low current densities and as a laser at high current densities (see figure 9.13). The

---

[*] The units for α used here are customary, but slightly unconventional. Normally, physicists use the neper/m as the unit for amplitude attenuation of an electric or magnetic field vector, so here α is twice as large as it would be in nepers/m; in other words, if α_t were 500 m^{-1}, that would be the same as 250 neper/m. The electrical engineer's unit for attenuation is the dB, which is 10log(Power Ratio) or 20log(V/V_0) etc. A neper is 8.686 dB, so 1 'laser' unit of attenuation is 4.343 dB.

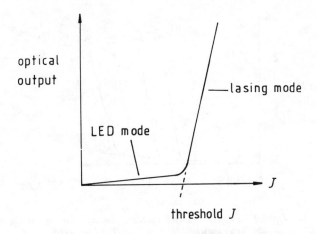

Figure 9.13 *Light output against current density for a p–n junction*

spectrum of the emitted light changes from being relatively broad during spontaneous emission to being very sharp (though not as sharp as most other types of laser) during stimulated emission. Defining Q as $\lambda_{max}/\Delta\lambda$, where λ_{max} is the wavelength of maximum output and $\Delta\lambda$ is the half power width, then for a normal LED Q is about 30; a superluminescent LED has a Q of about 300 and a junction laser has a Q of about 3000.

As the current density increases beyond the threshold level, very large increases in light output occur. Superluminescent diodes operate in the transitional region between LED and laser modes of operation. In the lasing region the output power can be written

$$P = P_0(J/J_{th} - 1)$$

where P_0 is a constant and J_{th} is the threshold current. In a homojunction laser $J_{th} \approx 500$ MA/m^2, while in a double heterojunction laser $J_{th} \approx 10$ MA/m^2, at 300 K. J_{th} increases exponentially with temperature. Typical output curves are shown in figure 9.14 for a stripe-contact, double-heterojunction laser, and it can be seen that the output at a given temperature is highly linear above J_{th}, an important consideration for modulation. Stripe-contact lasers (see figure 9.15) were developed because of output instabilities in other kinds, caused by spatial fluctuations in the lasing region. The stripe-contact is formed by proton bombardment of the areas outside the surface stripe to form high resistivity material. Current is forced into the unbombarded stripe and lasing is confined to the junction region under it.

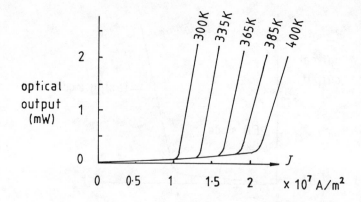

Figure 9.14 *Optical output against current density for a stripe-contact, double-heterojunction laser at from 300 K to 400 K*

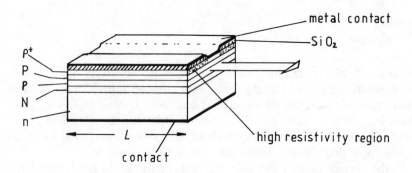

Figure 9.15 *A stripe-contact, double heterojunction laser*

Cavity Modes

In the laser there must be an integral number of half-wavelengths between the reflecting ends or the light wave will not be reinforced on reflection (the laser forms what is called a *Fabry–Perot cavity*). Now the light wavelength in the laser is λ/n, where λ is the vacuum wavelength and n the RI, so this condition is

$$m\lambda/2n = L$$

or

$$\lambda = 2nL/m$$

m is an integer. Ignoring the dependence of n on λ, we can differentiate λ with respect to m and get

$$d\lambda/dm = -2nL/m^2$$

Substituting for m $(= 2nL/\lambda)$ gives

$$d\lambda/dm = -\lambda^2/2nL \qquad (9.5)$$

The separation between allowable longitudinal modes, $\Delta\lambda$, is then $-\lambda^2/2nL$, which for $\lambda = 1.55$ μm, $L = 350$ μm and $n = 3.5$, is 1 nm. The output consists of a series of lines with a Gaussian intensity distribution grouped about the wavelength of maximum emission as in figure 9.16.

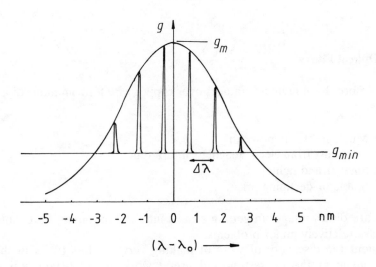

Figure 9.16 *Output from a solid-state laser*

For instance, a laser 500 μm long has maximum gain at 1.3 μm and has a RI of 3.55. The end reflections are 'natural'. If the gain is a Gaussian distribution with a maximum, $g_m = 4000$ m^{-1} and a spectral width of 20 nm at half maximum gain. How many longitudinal modes are operational if $\alpha_s = 800$ m^{-1}?

For the laser to emit light $g \geqslant \alpha_s + (1/2L)\ln(1/R^2)$. R is given by (9.3) as 0.314, so the minimum gain is 3117 m^{-1}. g is given by

$$g = g_m\exp[-(\lambda - \lambda_0)^2/2\sigma^2]$$

where $\lambda_0 = 1.3$ μm and σ is to be found. When $\lambda = \lambda_0 \pm 10$ nm, $g = \frac{1}{2}g_m$, so

$$0.5 = \exp(- 10^2/2\sigma^2)$$

and $\sigma = 8.49$ nm. Now g_{min}/g_m is 3117/4000, giving

$$3117/4000 = \exp[- (\lambda - \lambda_0)^2/2/8.49^2]$$

which means $(\lambda - \lambda_0) = \pm 6$ nm. Now the separation of modes is found from (9.5) to be 0.476 nm, so there are $2 \times 6/0.476 = 25$ modes.

It is possible to get monomodal operation in solid-state lasers by varying the RI of the laser's optical waveguide sinusoidally, which causes constructive interference essentially of only one mode. These *distributed feedback* lasers can be used in integrated optical systems and have greater temperature stability of the wavelength than normal lasers.

9.3 Optical Fibres

Optical fibres have great advantages over copper wire for communications purposes:

- Much greater bandwidth
- Immunity from electromagnetic interference
- Smaller and lighter
- Stable or declining price.

There are disadvantages in getting signals into and in joining fibres, but these are relatively minor problems.

Without the discovery of ways of making very low-loss fibres in the 1970s, much of the interesting and useful work on solid-state optical sources and detectors would have lacked impetus. Losses at 1.3 μm in silica fibres have now reached the theoretical minimum of about 0.3 dB/km. To put that into perspective, consider a window-pane 3 mm thick with absorption losses of 0.5 per cent. We do not notice this loss, which works out as 0.02 dB for 3 mm, 7 dB/m, or 7000 dB/km, but for optical communications window glass is opaque above a thickness of about 10 m. Though absorption and other losses must not be ignored, a more important consideration with optical fibres is bandwidth, which is largely determined by dispersion.

Dispersion is the term given to phenomena which tend to spread out an optical signal over time and frequency. In a PCM system, the energy in a light pulse which is 'on' is spread out so that if 'on' pulses are too close together they cannot be separated. There are two chief sources of dispersion in optical fibres: modal (or *inter*modal) dispersion and *intra*-

modal dispersion (made up of material and waveguide dispersions). We shall deal first with modal dispersion in two types of fibre in turn.

9.3.1 Step-index Fibres

A step-index fibre consists of a cylinder of material, RI = n_1, which is clad with a material of slightly lower RI, n_2. Modal dispersion is much the most important source of dispersion in these fibres and arises because there are many different paths down the fibre for light rays. Figure 9.17 shows several modes in a fibre. There will be an angle of incidence at which the ray will leave the fibre, that is when $\vartheta_i < \vartheta_c$, and such modes will not propagate. A thick fibre will have more modes than a thin one (see problem 9.5).

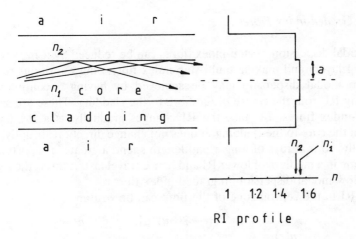

Figure 9.17 *Modes in a step-index fibre*

The energy of the light ray, and hence its signal content, moves with the group velocity of the ray, v_g, so the group delay in a fibre of length, L, for this ray will be $\tau = L/v_g$. The dispersion in arrival times because of multimode propagation will be the difference between the longest and shortest group delays. It turns out that the longest group delay is for rays which travel entirely in the central region for which $v_{g1} \approx c/n_1$, and the shortest group delay is for *evanescent* rays which travel in the cladding with group velocity, $v_{g2} \approx c/n_2$. The dispersion is then

$$\Delta\tau \approx L/v_{g1} - L/v_{g2} \approx L(n_1 - n_2)/c \qquad (9.6)$$

For a fibre 1 km long, with $n_1 = 1.52$ and $n_2 = 1.50$, the modal dispersion works out to 67 ns. In fibres rather longer than 1 km, the dispersion given by equation (9.6) is too great, because defects and kinks allow slower modes to catch up so that

$$\Delta\tau \approx \Delta\tau_1 \sqrt{L} \tag{9.7}$$

where $\Delta\tau_1$ is the dispersion for 1 km according to equation (9.6).

The maximum bit-rate, R, for PCM signals is given by

$$R \approx (4\Delta\tau)^{-1}$$

while the bandwidth for AM signals is roughly $R/2$. Thus over a 50 km length of the fibre in this example, $\Delta\tau \approx 67 \times \sqrt{50} = 474$ ns, and $R \approx 530$ kbits/s, a modest bit-rate. Monomodal step-index fibres can be made to eliminate modal dispersion, but the rather small core diameter (see problem 9.5) makes joining and signal insertion much more difficult. A better approach is the use of a RI which varies in the core material.

9.3.2   Graded-index Fibres

The modal dispersion of step-index fibres can be reduced by using several cladding layers and making multi-step-index fibres, but enormous reductions in modal dispersion may be achieved by having a continuously changing RI from the centre of the fibre to the cladding. These are called graded-index fibres. Because the RI changes gradually as a ray passes through the core to the cladding, it does not change direction abruptly, but gradually, so that rays of longer pathlength spend a greater proportion of their time in a medium of lower RI and hence travel faster: this is the secret of the low modal dispersion in graded index fibres.

The RI in the central zone of the fibre can be written

$$n(r) = \begin{cases} n_1 \sqrt{[1 - 2(r/a)^\alpha \Delta]}, & r < a \\ \\ n_2, & r > a \end{cases}$$

where r is the distance from the centre of the fibre, a is the radius of the core, $\Delta = (n_1 - n_2)/n_1$.[*] n_1 is the RI at the centre, n_2 is the RI of the cladding and α is the grading parameter ($\alpha = 1$ for linear grading, $\alpha = 2$ for parabolic and $\alpha = \infty$ for a step-index fibre). Figure 9.18 shows $n(r)$ for various values of α. The optimal value for α is given by

$$\alpha_{opt} = 2(1 - \Delta)$$

that is, a little less than two. For optimal α the dispersion is given by

$$\Delta\tau_{min} = Ln_1\Delta^2/8c \tag{9.8}$$

---

[*] Because this usage – though conventional in optoelectronics – is confusing, Δ is placed at the end of a numerator, of a denominator or of a formula in this text.

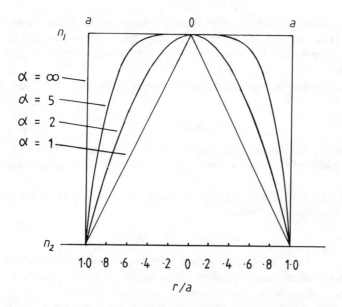

Figure 9.18 *RI profiles for some values of the grading parameter,* α

When α is a little removed from its optimal value, the dispersion is

$$\Delta\tau = L(n_1 - n_2) \, | \, \alpha - \alpha_{opt} \, | \, /c(\alpha + 2) \qquad (9.9)$$

Suppose $n_1 = 1.5$ and $\Delta = 0.01$, then $\alpha_{opt} = 1.98$, $\Delta\tau_{min}/L = 63$ ps/km, and if $\alpha = 2$, $\Delta\tau/L = 0.25$ ns/km. The dispersion has a very sharp minimum at α_{opt}, which is easily missed in manufacturing the fibre.

The number of modes supported by the graded-index is approximately given by

$$N \approx (k_1 a)^2 \alpha\Delta/(\alpha + 2) \qquad (9.10)$$

where $k_1 = 2\pi n_1/\lambda$. When $\alpha \approx 2$, N is half that for a step-index fibre, which leads to half the energy input to it compared with the step-index fibre, as the energy in each propagating mode is the same.

9.3.3 Intramodal Dispersion

Intramodal dispersion is made up of material dispersion and waveguide dispersion. The former is caused by the RI being a function of wavelength, and would occur regardless of the fibre dimensions. Waveguide dispersion is caused by variation in the propagation constants of the waveguide with frequency. Since waveguide dispersion is proportional to $1/a^2$, where a is the fibre radius, it is most important for monomodal fibres. Waveguide dispersion is negative while material dispersion is positive, so in these fibres it is possible to eliminate intramodal dispersion at a particular wavelength.

In silica fibres, the material dispersion is negligible at $\lambda = 1.3$ μm, and is roughly

$$\Delta\tau \leqslant 30\Delta\lambda L \qquad \text{μs}$$

elsewhere in the wavelength region from 1 to 2 μm, where $\Delta\lambda$ is the spectral width of the light source. There is no $\sqrt{L}$ dependence for 'long' fibres.

Intermodal and intramodal dispersions are not directly additive, but must be combined as follows

$$\Delta\tau = \sqrt{(\Delta\tau_a^2 + \Delta\tau_e^2)}$$

where $\Delta\tau_a$ and $\Delta\tau_e$ are the respective intramodal and (inter)modal dispersions for the entire length of fibre. Intramodal dispersion can be reduced by using a source of narrow spectral range, such as a laser.

For instance, a step-index fibre 2 km long has a core with $n_1 = 1.52$ and a cladding with $n_2 = 1.51$. It is operated at 1 μm with a LED source of spectral width 60 nm. What is the maximum bit-rate possible?

The modal dispersion will given by (9.6) and (9.7) as $\Delta\tau_e = \sqrt{2} \times 1000 \times 0.01/3 \times 10^8 = 47$ ns. The intramodal dispersion can be taken to be due solely to material dispersion, which will be of the order of $30\Delta\lambda L$ μs since the wavelength is some way from 1.3 μm. So $\Delta\tau_a = 30 \times 2000 \times 60 \times 10^{-9}$ μs $= 3.6$ ns, and $\Delta\tau = \sqrt{(47^2 + 3.6^2)} = 47$ ns: the material dispersion is completely swamped by the modal dispersion. The maximum bit-rate, R, is $1/4\Delta\tau = 5.3$ Mbit/s. If we now consider a graded-index fibre, with optimal grading index, taking $n_1 = 1.52$ and $n_2 = 1.51$ as before, then the modal dispersion is at its minimum value, $\Delta\tau_e = 1.52 \times (0.01/1.52)^2 \times 1000/(8 \times 3 \times 10^8) = 27.4$ ps/km, from equation (9.8), or $27.4\sqrt{2} = 39$ ps altogether. The material dispersion will remain at 3.6 ns, so the total dispersion is 3.6 ns and $R = 70$ Mbit/s. In this case the modal dispersion is swamped by the material dispersion. It is unlikely that α will be quite at its optimal value of 1.98, but $|\alpha - \alpha_{opt}|$ will be about 0.05, so that, from (9.9), $\Delta\tau_e = 0.59$ ns, $\Delta\tau = 3.64$ ns and $R = 69$ Mbit/s. Only when the material and modal dispersions are comparable does one not completely overshadow the other.

9.3.4 *Attenuation in Optical Fibres*

At first, attenuation was the most important obstacle to the use of optical fibres, since about 1000 dB/km was the best obtainable in the 1960s. Subsequent improvements have resulted in an attenuation coefficient for silica fibres as shown in figure 9.19, which is very close to the theoretical minimum. The long-wavelength limit in the infra-red is caused by electronic lattice transitions, especially the strong absorption around 10 μm, which has a large 'tail' down to 1.5 μm. This limit is an intrinsic property of silica, and can only be overcome by changing to other materials, such as alkali halides or fluoride glasses, which have no infra-red absorption up to longer wavelengths than 10 μm. The absorption peak at 1.4 μm is due to hydroxyl ions in the silica, which cannot be eliminated. At short wavelength the limit is set by Rayleigh scattering. Silica is a glass and as such consists of small volumes in which the atoms (about a dozen in number) are correctly positioned, joined to other such groups of atoms in a haphazard way. These groups scatter light because of micro-variations in the RI. The scattering losses go as λ^{-4}, amounting to 4.5 dB/km at 0.6 μm and 0.3 dB/km at 1.2 μm. With reasonable sources and receivers, losses of 0.3 dB/km mean that repeaters are not necessary in fibres up to 100 km long. Coupling and bending losses are more significant in most cases.

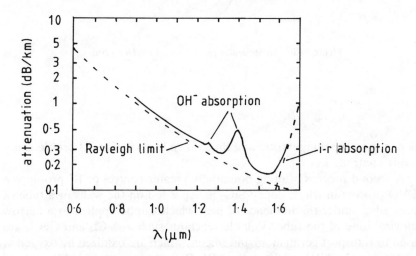

Figure 9.19 *Attenuation against wavelength for silica fibres*

9.3.5 The Manufacture of Optical Fibres

The basic process is simple in principle. The core material is melted in a container with a hole in the bottom which is centrally placed in another vessel containing the molten cladding material, as in figure 9.20. Graded-index fibres may be obtained by adding dopants such as GeO_2 to the molten SiO_2 which diffuse into the central core by an amount which

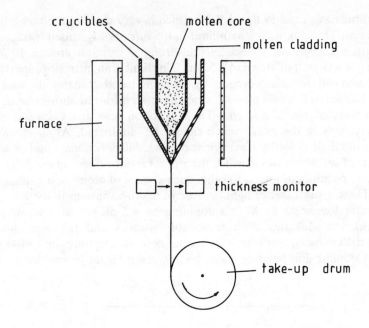

Figure 9.20   *An apparatus for producing cladded fibres*

depends on the temperature and time taken to cool. The RI profile is not easily controlled.

A second process offering potentially greater control of RI profile is a CVD process in which doped SiO_2 is deposited on the walls of a tube of pure silica under the influence of heat which can be applied to a narrow moving zone of the tube. Volatile reactants such as $SiCl_4$ and $GeCl_4$ are used to transport germanium and silicon, which are oxidized by oxygen at 1200°C to $(Ge, Si)O_2$ of the desired RI. Repeated passes with differing gas compositions can be made to give the desired profile. The finished tube is then heated to beyond its softening point (about 1800°C) so that it collapses into a solid rod, which can then be drawn down to the desired diameter. This is largely determined by the temperature of the undrawn fibre (through its very pronounced influence on viscosity), though the draw-rate is also important and can be controlled by feeding back thickness information to the drum motor. Deposition can also be carried out on the outside of a rod or tube in much the same way to produce the required index gradient.

9.4 Signal Detectors

The photodetectors used in a fibre-optic communications system are usually p–n junction devices like the PIN diode and the avalanche photodiode (APD). The diodes are used in the third quadrant of their *V–I* characteristic, that is, in the region of reverse bias and reverse current flow, sometimes called third-quadrant mode (TQM) as opposed to first-quadrant mode (FQM), or forward biased. In a condition of no illumination, there is still a small current which will flow under reverse bias called the *dark current*; in photodiodes this is greatly increased by the presence of electron–hole pairs created by the incidence of a photon with a wavelength less than λ_g, the gap-energy wavelength. The ratio of the number of electron–hole pairs created to the number of incident photons is the quantum efficiency, η. If i_{ph} is the photon current (due solely to electrons, assuming, $\mu_n \gg \mu_p$), then the number of electron–hole pairs created per second is i_{ph}/e, where i_{ph} is the photocurrent. As each requires energy hf, the power required for the photocurrent is $i_{ph}hf/e$. If the input light power is P_i

$$\eta P_i = i_{ph}hf/e \qquad (9.11)$$

The *responsivity*, $\mathcal{R}_0$, is defined as i_{ph}/P_i, so that

$$\mathcal{R}_0 = \eta e/hf = \eta\lambda/1.24$$

when λ is in μm. Thus the ideal responsivity is a straight line from the origin ($\lambda = 0$ nm) to λ_g, where it falls abruptly to zero, as λ must be less than λ_g for photon production. Responsivity has units of A/W.

9.4.1 PIN Diodes

Figure 9.21 shows a PIN diode in cross-section. It derives its name from the fact that the doping sequence of the layers is p–i–n, where i stands for intrinsic. The surface on which the light is incident is coated with an antireflectant layer so that all the photons enter the diode. The fraction of photons absorbed in a thickness of diode material, x, is $1 - \exp(-\alpha x)$, where α is the absorption coefficient. If $\alpha \approx 10^4 \ \text{m}^{-1}$, as in silicon at $\lambda = 1 \ \mu$m then about 100 μm thickness is needed to absorb 63 per cent. In a straightforward p–n junction diode, the width of the depletion region in, or very near to, which the photons must be absorbed, is proportional to $\sqrt{V}$ (see chapter 4 for the underlying theory of p–n junctions). Thus large reverse bias is needed for good efficiency.

In the PIN diode, the depletion region includes the whole of the intrinsic layer, which can therefore be made as large as necessary. Conversely, since the carriers must be generated within a diffusion length, $\sqrt{[D\tau]}$, of the

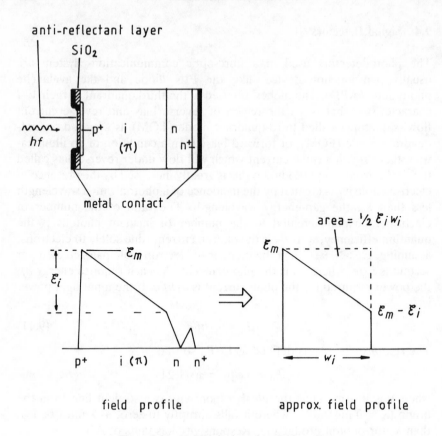

Figure 9.21 *A PIN diode and its field profile*

junction – or they will recombine before detection – the p+ region must be very thin, since absorption coefficients for semiconductors for wavelengths less than λ_g become very large. Figure 9.22 shows some absorption coefficients on a log scale as a function of photon energy, from which we see that α rises extremely rapidly when $E_{ph} > E_g$ (or $\lambda < \lambda_g$), for direct-gap materials such as GaAs, but rather less rapidly for indirect-gap materials like silicon. Thus silicon is a material well suited for wide-spectrum photodetection, working efficiently over the wavelength range 0.4 to 1.1 μm, with peak responsivity around 0.95 μm.

As an example, a photodiode has a depletion region 20 μm thick, and the thickness of material a photon must cross before reaching the depletion region is 5 μm. The surface of the diode is not treated and the RI is 3.5, while $\alpha = 5 \times 10^4$ m^{-1}. What is the efficiency?

First, the transmissivity is given by $1 - (n_1 - n_2)^2/(n_1 + n_2)^2$, which is $1 - (3.5 - 1)^2/(3.5 + 1)^2$, or 0.69 – 69 per cent of the incident photons

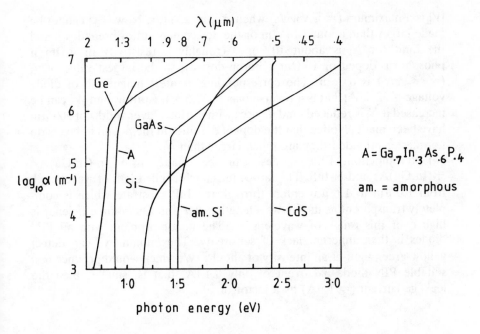

Figure 9.22 *Absorption coefficients against photon energy for Si, Ge, GaAs, $Ga_{0.7}In_{0.3}As_{0.4}P_{0.6}$, CdS and amorphous Si*

enter the diode. The fraction of these photons transmitted through the first layer is $\exp(-5 \times 10^4 \times 5 \times 10^{-6}) = 0.78$. The fraction absorbed in the depletion region is $1 - \exp(-5 \times 10^4 \times 20 \times 10^{-6}) = 0.63$, so $\eta = 0.69 \times 0.78 \times 0.63 = 34$ per cent. An antireflectant coating, a thinner top layer and a wider depletion region could boost this figure to 80 per cent.

PIN diodes can be made with very large depletion regions by simply increasing the thickness of the intrinsic layer, while the top p- (or n-) layer can be made very thin. Figure 9.21 shows the electric field profile across a PIN diode. (We use in what follows Greek letters for material which is *nearly* intrinsic, $\pi \equiv p$ and $\nu \equiv n$.) The field gradient, $d\mathscr{E}/dx$, is eN/ϵ, where N is the dopant density and ϵ is the permittivity, $\epsilon_0\epsilon_r$, and as N_A, $N_D \gg N_i$, the field changes abruptly at the p–π and π–n junctions (N_A, N_D and N_i are the respective dopant densities of the p-, n- and π-layers). The (reverse) voltage across the diode is just the area under the field profile graph (see figure 9.21), that is, $|\mathscr{E}_m|w_i - \frac{1}{2}|\mathscr{E}_i|w_i$, where $|\mathscr{E}_m|$ is the size of the maximum field in the diode. $|\mathscr{E}_i|$ is the size of the field across the intrinsic layer, width w_i, which is $eN_iw_i\sqrt{\epsilon}$, so

$$|V| \approx |\mathscr{E}_m|w_i - \tfrac{1}{2}eN_iw_i^2/\epsilon$$

$|V|$ is a maximum ($= \frac{1}{2}eN_iw^2_i/\epsilon$) when $|\mathscr{E}_m| = eN_iw_i/\epsilon$. Now $|\mathscr{E}_m|$ cannot be made larger than about 1 MV/m or the device may suffer breakdown and the limit for N_i is about 10^{19} m^{-3} (roughly a resistivity of 4 Ωm if phosphorus-doped or 10 Ωm if boron-doped), so the largest value of w_i ($= \epsilon\mathscr{E}_m/eN_i$) is 65 μm. The corresponding greatest magnitude of diode voltage is 32.5 V (that is a reverse bias of 32.5 V). Both w_i and $|V|$ can be increased if N_i is reduced, and devices with intrinsic layer widths of 500 μm have been made. Notice that the dopant densities in the p- and n- layers do not matter provided they are much larger than N_i.

Heterojunction PIN diodes can be made from p-(In,Ga)As, π-(In,Ga)As and N-InP. The lattice mismatch is nil. Light of wavelength from 1.0 μm to 1.6 μm enters through the InP substrate which is completely transparent as its gap wavelength is 0.9 μm. Quantum efficiency is high over this range of wavelengths. The major problem with all PIN diodes is their inherent lack of sensitivity. The human eye can detect yellow-green light at an intensity of about $\frac{1}{2}$ fW, which would produce in a suitable PIN diode a current of only 0.1 fA, that is, smaller than the leakage current ($\approx$ 1 nA) by a factor of 10^7.

9.4.2 Avalanche Photodiodes

The photocurrents produced in PIN diodes are small – the device has no amplification – whereas an avalanche photodiode (APD) offers carrier multiplication by utilizing the avalanche effect. If the current in the APD is i_{apd}, then the current multiplication, M is just i_{apd}/i_{ph}, so that $\mathscr{R}_{apd} = M\mathscr{R}_0$. A high field is required in the diode to get avalanche multiplication under reverse bias, so a heavily-doped n$^+$-layer is used at the light input side, followed by a p-layer, in which the avalanche effect occurs, and then a π-layer across which most of the voltage is dropped, though the field gradient is small. A p$^+$ layer completes the structure, which is shown in figure 9.23.

The current amplification depends exponentially on the width of the p-layer, w_a, so that $i_{apd} = i_{ph}\exp(\alpha w_a)$, where α is the ionization coefficient, and $M = \exp(\alpha w_a)$. It turns out that M varies very rapidly with voltages near the breakdown voltage when the ionization coefficients for holes and electrons are nearly equal, but much less rapidly when electron ionization predominates, so that essentially only one type of carrier is present. The chief disadvantage of the APD is its high working voltage, 100–500 V, and its sensitivity to temperature. In addition, for stable gain, the high working voltage must be kept very constant. Because of these considerations, the gain is limited and M is about 10–100.

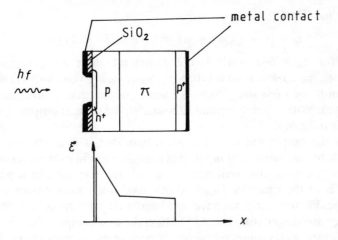

Figure 9.23 *An APD and its field profile*

9.4.3 Phototransistors

In a phototransistor the base current is the photocurrent produced by incident photons, so the base lead is not required. In figure 9.24 are shown the currents flowing in an npn phototransistor. The collector current, i_c, is made up of the reverse saturation current, i_s and the fraction of the emitter current that does not suffer minority carrier recombination in the base, αi_e, where α is the common-base current gain (≈ 0.99). Now since $i_c + i_b = i_e$ we have

$$i_s + \alpha i_e + i_{ph} = i_e$$

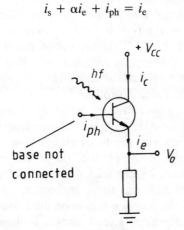

Figure 9.24 *Currents flowing in an n–p–n phototransistor*

since $i_b \equiv i_{ph}$. Thus

$$i_e = (i_s + i_{ph})/(1 - \alpha) = (i_s + i_{ph})(\beta + 1)$$

where β (or $h_{fe}, \equiv \alpha/[1 - \alpha]$) is the common-emitter current gain. If $i_{ph} = 0$, then the *dark current* is $i_s(\beta + 1)$. Since i_{ph} is given by (9.11) it will be of the order of a few μA. Whereas i_s will be of the order of nA, and can be neglected. With β being typically about 100, the current output, i_e, will be about a milliamp.

Though the output current of a phototransistor is β times that of a photodiode for the same light input, this improvement in gain comes at the expense of a comparable reduction in speed: the rise time of a photo-transistor is of the order of 10 μs, about 1000 times slower than a PIN photodiode. Photodarlingtons have also been used, with gains of 30 000 or so, but they are slower still. The APD offers the best compromise between the PIN diode and the phototransistor, though often a combined package of PIN diode and fast silicon op-amp is used.

9.5 The Solar Cell

Solar cells have become efficient and cheap enough to provide electrical power in MW. They are noise- and pollution-free, so their energy is at a premium. Unfortunately they occupy a lot of space and their use as power sources has been restricted to hot desert locations such as Arizona, Nevada and parts of California. Hot deserts are growing in area, however, so the potential of solar cells is growing. Because solar cells, unlike steam-driven power stations, do not work on a Carnot cycle, they do not suffer from the enormous loss of energy due to low thermal efficiency. (The Carnot efficiency is $\eta = (T_h - T_c)/T_h \approx 0.5$ for a power station, where T_h is the inlet steam temperature and T_c is the outlet steam temperature of the turbogenerators.) In tonnage terms, solar cells may be the biggest con-sumer of semiconductor material already. The material used for large-area solar cells is amorphous silicon produced by the low-temperature decom-position of silane on a glass substrate. Solar cells operate in the same way as photodiodes: an incoming photon, of wavelength less than the gap wavelength, is absorbed and produces an electron–hole pair, delivering energy, eE_g, to the cell. If the photon has greater energy than eE_g, the excess is wasted as heat. Figure 9.25 shows a solar cell in cross-section. Incoming light passes through an antireflectant coating and generates electron–hole pairs in the junction region on the low-resistivity p-side. The solar cell has the equivalent circuit shown in figure 9.26(a) and the corresponding I–V characteristic shown in figure 9.26(b). Solar cells operate in the fourth-quadrant mode, that is, the current supplied is negative (it is opposite in sign to the diode current I_d, which is

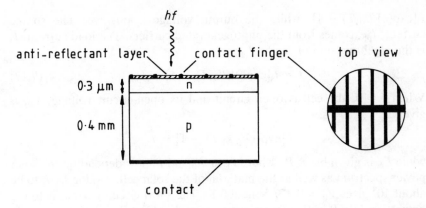

Figure 9.25 *A solar cell*

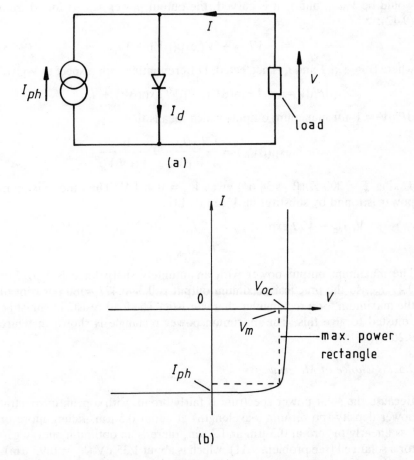

Figure 9.26 *(a) The equivalent circuit of a solar cell; (b) the I(V) characteristic of a solar cell*

$I_s[\exp(eV/k_BT) - 1]$, while the output voltage is positive. The source current, I_{ph}, comes from the photogenerated carriers. The load current, I, is the algebraic sum of I_d and I_{ph}:

$$I = I_s[\exp(eV/k_BT) - 1] - I_{ph} \qquad (9.12)$$

When $I = 0$, the cell is open circuit and its open-circuit voltage, V_0, is given by

$$I_s[\exp(eV_0/k_BT) - 1] = I_{ph}$$

while I_{ph} is given by $\mathcal{R}_0 P_i$, where both $\mathcal{R}_0$ and P_i now depend on the solar power spectrum as well as the material of the solar cell. Taking I_{ph}/I_s to be about 10^8 gives $V_0 = 0.477$ V at 300 K. The short-circuit current is found by putting $V = 0$ in (9.12), to give $I_{sc} = - I_{ph}$. If the knee of the curve in the fourth quadrant were infinitely sharp, then the maximum output power would be $V_0 I_{ph}$, but as it is curved, the output power can be found from (9.12):

$$P = - VI \approx - VI_s\exp(\beta V) + VI_{ph} \qquad (9.13)$$

where $\beta \equiv e/k_BT$ and I_s is neglected. Differentiating with respect to V gives

$$dP/dV = - I_s\exp(\beta V) - \beta VI_s\exp(\beta V) + I_{ph}$$

$dP/dV = 0$ for maximum output, which means that

$$\exp(\beta V_m) = \frac{I_{ph}/I_s}{1 + \beta V_m} = \frac{10^8}{1 + \beta V_m}$$

Taking $T = 300$ K ($\beta = 38.64$) gives $V_m = 0.404$ V. Then the maximum power is found by substituting V_m in (9.13);

$$V_m I_{ph} - V_m I_s\exp(\beta V_m) = V_m I_{ph} - 10^{-8}V_m I_{ph}\exp(\beta V_m)$$

$$= 0.38 I_{ph}$$

The maximum output power with an infinitely sharp knee is $V_0 I_{ph}$, or $0.477 I_{ph}$, so the practical maximum output is $0.38/0.477 \approx 80$ per cent of the maximum for a rectangular characteristic. The load resistance must be adjusted to give this. The maximum power rectangle is shown in figure 9.26(b).

9.5.1 Choice of Material

Because the solar power spectrum is fairly broad, with a peak in spectral power density (power/unit wavelength) at about 0.5 µm, falling more or less linearly to zero at 0.3 µm and 2 µm, there is an optimum energy gap for a solar cell (see problem 9.11), which is about 1.35 eV ($\lambda_g = 0.92$ µm), so silicon is quite a good match ($E_g = 1.1$ eV). Amorphous silicon has a rather larger band gap (about 1.5 eV), so is perhaps an even better match.

The maximum output from an ideal bandgap semiconductor is about 30 per cent of the solar power input. In addition, the window passes only 90 per cent of the incoming radiation, the quantum efficiency is about 80 per cent and the load matching (maximum power rectangle) is only 80 per cent, so the overall efficiency is $0.3 \times 0.9 \times 0.8 \times 0.8 = 17$ per cent. A single-crystal silicon solar cell typically has an efficiency of about 13 per cent.

9.6 Display

Displays can be divided into active (light-emitting) and passive (which modify ambient illumination). LEDs and CRTs fall into the first category, while LCDs fall into the second. An alternative classification is to consider status and alphanumeric displays separately from graphical. In this case, LEDs and LCDs fall into the first category and CRTs the second. There is no doubt that the CRT is the most important display device at present.

9.6.1   *The Cathode Ray Tube (CRT)*

The best display for ease of reading and writing and cost/bit is the CRT, and will remain the CRT until at least the year 2000. Though its demise has been predicted for some thirty years, it still occupies the most important place in the display field. What can be the reason for this? Is it not an antiquated vacuum-tube device, whose origins lie far back in technological time? Why has it not been swept away, like vacuum-tube diodes and amplifiers, by the tide of solid-state technology? The answer is compounded of two factors: addressing and power consumption. In a solid-state display the addressing must be done by a matrix of conductors, while in the CRT addressing by deflecting an electron beam is simple and very flexible. Addressing also includes the display of colour, which is very difficult for a solid-state display, but easily achieved by a CRT. Power consumption is not high in a CRT because the conversion of electron-beam energy to light in a phosphor is relatively efficient (about 25 per cent) compared with, say, a LED (about 1 per cent).

Colour displays are achieved on a CRT by putting down a pattern of red, green and blue phosphor dots on the back of the screen which are then addressed either by three separate electron beams (as in the shadow-mask TV tube) or by a single, colour-modulated beam. Materials research in the CRT field has concentrated on phosphors and the advent of colour television has resulted in considerable improvements in phosphors, the most notable of which has been the great increase in brightness of the red phosphor which is now based on Y_2O_2S doped with Eu^{3+} and Tb^{3+}. These phosphor materials were suggested to researchers by the infra-red laser which used yttrium iron garnet doped with neodymium (YIG:Nd).

The rare earth (RE) ion (Eu^{3+} or Tb^{3+}) is excited by an electron from the gun so that one of its 4f-electrons is excited into a higher state from which it decays by a path involving the emission of a photon at essentially one wavelength. The older phosphors did not use RE ions and the interaction between emitting atom and host lattice resulted in a broad output spectrum. The 4f-electrons of all the REs lie deeply in the electronic structure, screened from lattice influences by the 5s and 5p electrons, so giving reasonable monochromaticity. However, the green and blue phosphors are still based on non-RE-doped II–VI compounds such as (Cd, Zn)S:Cu (green) and ZnS:Ag (blue).

Pigmented phosphors have also been developed which absorb ambient light and not the phosphor-emitted light. These pigments can be attached to the surface of the phosphor grain and improve contrast. Narrow-band phosphors have the advantage that contrast-enhancing filters can be used efficiently with them, whereas broad-band phosphors do not permit such enhancement (because too much of their output would be absorbed).

Further areas of phosphor research are into persistence and resistance to damage from the electron beam. Again, YAG:RE materials seem to have some potential for damage resistance.

9.6.2 *Liquid Crystal Displays*

LCDs are the only widely used, passive, display device. They have the great advantage of consuming very little power: a 6-digit display with 12.7 mm digits will consume about 100 μW. This is chiefly because they are passive; by way of contrast, a 6-digit, 7-segment, 12.7 mm red LED display will consume about 0.75 W when all the digits are lit – 750 times as much power as the LCD. For battery-powered devices, the difference is decisive.

LCDs are based on the fact that long, polar, organic molecules can be aligned by an electric field and in this condition they do not reflect much light. The liquid-crystal state is found near the freezing point of the liquid, and is a transitional phase between solid and liquid with a restricted temperature range, typically from 0°C to 50°C. In some instances the ordering is helical in character with the pitch of the helix being dependent on the temperature. In these cases the crystal reflects light when the pitch has an integral number of wavelengths and not otherwise, so white light is reflected back as coloured light and passive, self-reading, digital thermometers can be made for home aquaria and similar uses.

In an addressable LCD display the liquid is in the form of a layer about 10 μm thick held between polaroid plates. The molecules are induced by surface treatment of the polaroid to line up in the polarizing direction. If the polarizing directions are at right angles as in figure 9.27(a), then incident light entering the liquid crystal is rotated by the liquid crystal

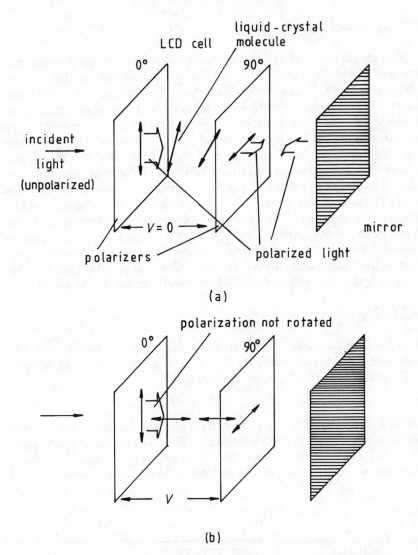

Figure 9.27   *(a) A liquid-crystal cell with no voltage applied; (b) a liquid-crystal cell with working voltage applied*

molecules, passes through the back polarizer and is reflected by the mirror, to return by the same path. When an electric field is applied between the plates, the liquid crystal molecules line up as in figure 9.27(b) so that the incoming polarized beam is not rotated and cannot pass through the rear polarizer to the mirror, so that little is reflected back. The aligning field is of the order of $\frac{1}{2}$ MV/m, so quite low voltages serve to operate the device.

9.7 Integrated Optics?

Although the idea of integrated optical signal-processing systems was mooted as long ago as 1965 (in *Optical and Electro-optical Information Processing* by D. B. Anderson, MIT Press, Cambridge, MA, USA), not much has been achieved outside the laboratory. The idea is simple: do as much signal-processing on the optical signal without first converting it to an electrical signal. It has not progressed much because of the tremendous advances made in the speed of digital signal processing ICs made from silicon and a concomitant price reduction. The balance of advantage has shifted progressively towards electrical signal-processing in the last few years because of the impact of VLSI technology. In a different field – underwater acoustics – the advantages of sound beam-shaping by lenses have been greatly reduced by the advent of fast and cheap signal-processing chips as well as the use of microprocessors to control the input signals to multi-channel arrays. However, it is still possible that a cheap, fast, optical switch will be found which could displace the transistor.

9.7.1 The Optical Switch

This consists essentially of a waveguide split into two halves, one with electrodes, which can modify the RI of the material between, as in figure 9.28. The change of RI can be expressed as

$$\Delta(1/\epsilon_r) = \Delta(1/n^2) = r\mathscr{E} \tag{9.14}$$

where n is the RI, $\mathscr{E}$ is the electric field and r is the *linear electro-optic coefficient*. This change in RI with electric field is called the *Pockels effect*. There is a quadratic field term (which gives rise to the optical Kerr effect) also, but we shall neglect it. If a suitable field is applied to the material

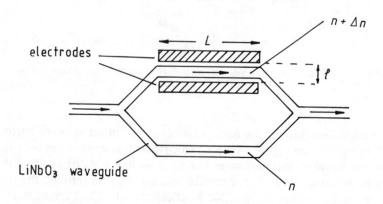

Figure 9.28 *An optical switch*

between the electrodes, then its RI is changed and the phase of the optical signal is altered as it traverses the waveguide. If this phase change is adjusted to 180°, then it will cancel the other half of the signal in the switch, and the beam is switched off. It can be shown (see problem 9.12) that if the separation of the electrodes is l and the wavelength of the light is λ, while the path length between the electrodes is L, that the phase shift is given by

$$\Delta\phi = \pi r n^3 V L / l \lambda \qquad (9.15)$$

In the case of $LiNbO_3$ (lithium niobate, of structure akin to perovskite), $r \approx 31$ pm/V, while $n \approx 2.3$, so if $L = 100$ µm, $\lambda = 0.6$ µm and $l = 10$ µm, then $V = 80$ volts, a fairly modest voltage. The switching time is about 100 ps.

Problems

9.1. If the eye's response to light (y) is a Gaussian curve ($y = y_m \exp[- (x - x_0)^2/2\sigma^2]$, y_m is the max. response) with its maximum at $x_0 = 555$ nm and $\sigma = 42$ nm, what will be its response (relative to its maximum response) at 400 nm, 700 nm and 800 nm? Use figures 9.1 and 9.2 to decide on the composition of a Ga(As, P) LED material, not doped with nitrogen, which will have the greatest apparent brightness for a given input power.
   [Ans. 0.11 per cent, 0.26 per cent, 4×10^{-6} per cent]

9.2. Show that the spectral width of a LED is given by

$$\Delta\lambda \approx 2.4 k_B T \lambda^2$$

where λ is in µm and $k_B T$ is in eV. What will be the Q of a GaAs LED and of a red-emitting Ga(As, P) LED at 77 K and at 300 K? $[Q = \lambda_{max}/\Delta\lambda.]$
   [Ans. 73, 19; 103, 26]

9.3. Consider the double interface between media of RIs n_1, n_2 and n_3, so that the transmission coefficient of the combination is $T = T_1 T_2$, where T_1 is the transmission coefficient at the interface between materials 1 and 2 and T_2 that between materials 2 and 3. T and T_2 are given, with necessary changes, by equation (9.3). Taking normal incidence only, show that maximum transmission is obtained when $n_2 = \sqrt{(n_1 n_3)}$, and medium 2 is then a matching layer. If $n_1 = 3.5$ and $n_3 = 1$, how much improvement in transmission is obtained with a matching layer compared with none (that is, when $n_2 = 1$)?
   [Ans. 19 per cent]

9.4. Use figure 9.6 to determine the average change in bandgap and lattice parameter when In is replaced by Ga in InAs and InP. Then find similarly the same changes when P is replaced by As in GaP and

InP. Hence determine the quaternary composition which will have the same lattice parameter as InP and will produce light of wavelength 1.3 μm.

9.5. What will be the critical angle for a ray in a step-index fibre for which $n_1 = 1.53$ and which has a cladding whose RI is 2.5 per cent less? When the number of modes propagating, N, is large, it is given by equation (9.10), with $\alpha = \infty$. What will N be for this fibre if its core radius is 20 μm and $\lambda = 1.3$ μm? Derive an equation for the maximum value for the radius, a_{max}, to ensure only one mode propagates.

[N is not 'large' in this case and a_{max} is too small by a factor of $1.2\sqrt{2}$.]

[*Ans. 77°; 547*]

9.6. A fibre-optic link is operated at a wavelength of 1.3 μm. Find the maximum bit-rate of a 25 km line made from (a) a multimodal step-index fibre, (b) a monomodal step-index fibre, (c) a graded index fibre of suboptimal grading such that $|\alpha - \alpha_{opt}| = 0.04$ and (d) an optimally graded fibre. [Take $n_1 = 1.47$ and $n_2 = 1.46$ in all cases and the source to be a LED. You must estimate the bandwidth at 300 K.] What would the bit rates be if the LED were operated at 77 K? What would the bit rates be with a laser source for which $\Delta\lambda = 1$ nm?

[Find the (inter)modal dispersion and assume the intramodal dispersion is $9\Delta\lambda L$ μs, or 9 ps/(nm-km), in all cases.]

[*Ans. LED at 300 K, (a) 1.5 Mb/s; (b), (c) and (d) 10.4 Mb/s. LED at 77 K, (a) 1.5 Mb/s, (b) 40 Mb/s, (c) 39 Mb/s, (d) 40 Mb/s. Laser, (a) 1.5 Mb/s, (b) 1.09 Gb/s, (c) 148 Mb/s, (d) 1.09 Gb/s.*]

9.7. Ignoring attenuation, what is the longest line that can be operated in each of the twelve cases of the previous question if it must carry 1000 speech channels?

[A speech channel requires a bit-rate of about 56 kbit/s.]

[*Ans. LED at 300 K, (a) 0.135 km, (b) 4.8 km, (c) 4.7 km, (d) 4.8 km. LED at 77 K, (a) 0.135 km, (b) 18.5 km, (c) 17.6 km, (d) 18.5 km. Laser, (a) 0.135 km, (b) 500 km, (c) 163 km, (d) 495 km*]

9.8. A PIN diode is made from germanium which has an intrinsic carrier concentration of 3×10^{19} m^{-3}. What will be the maximum width of the intrinsic region if the maximum field in the device is 10^6 V/m? If the absorption coefficient $\alpha = 10^4$ m^{-1} for a particular wavelength, what is the maximum possible quantum efficiency? By cooling the diode to 77 K, the intrinsic dopant concentration is reduced to 3×10^{18} m^{-3}. What is now the maximum quantum efficiency?

[ϵ_r for Ge is 16.]

[*Ans. 29.5 μm. 25.6 per cent. 94.8 per cent*]

9.9. Estimate the responsivity, $\mathscr{R}_0$, of a silicon PIN diode at $\lambda = 1$ μm, if $w_i = 100$ μm, $\alpha = 10^4$ m^{-1} and all the incident photons enter the intrinsic region. If the light power incident on the diode is 10 μW, what current will flow? If the diode capacitance is 40 pF and the load impedance is 50 Ω, what is the response time?
[*Ans. 0.51, 5.1 μA, 2 ns*]

9.10. A silicon APD is coated with an antireflectant layer. What should its refractive index be? If the width of the action region, w_a is 1.5 μm and the ionization coefficient is 2×10^6 m^{-1}, what is the current amplification factor, M? If the silicon has an absorption coefficient of 5×10^4 m^{-1} and the intrinsic layer is 25 μm wide, what will be the quantum efficiency? Estimate the responsivity at $\lambda = 1$ μm, and the current flowing when the light power incident on the diode is 10 μW.
[*Ans. 1.85, 20.1, 55 per cent, 8.9, 89 μA*]

9.11. Find the optimal bandgap for a solar cell material. What will be the greatest efficiency possible for the optimal material?
[Take the solar power/unit wavelength to be peaked at 0.5 μm falling linearly to zero at 0.3 μm and 2 μm. Remember that for the cell to absorb a photon $\lambda_{ph} < \lambda_g$ and the electrical energy from that absorption is eE_g.]
[*Ans. 71 per cent*]

9.12. Derive equation (9.15) from equation (9.14).

10

Superconductors

Kammerlingh Onnes coined the term *superconductivity* in 1911 to describe the state of a sample of mercury, whose electrical resistance suddenly became too small to measure at a temperature of about 4 K. The work Onnes was doing was a consequence of his earlier experimental achievement, the liquefaction of helium in 1908. Before helium was liquefied, there was no way of cooling a body much below the boiling point of hydrogen (about 20 K), far too high a temperature for any element to exhibit superconductivity. From 1911 until perhaps 1986, superconductivity has always required liquid helium for its practical attainment. And this requirement has stifled most large-scale applications for superconductors because using liquid helium is expensive – expensive in refrigeration energy and in its requirement for the most elaborate means of thermal insulation.

10.1 The Economics of Superconductivity

Refrigeration is cheap when the difference between the working temperature and that of the heat sink is not too great. A domestic refrigerator requires about 0.2 J to remove 1 J of heat from its interior. However, liquid helium boils at 4.2 K and 300 J must be expended to remove 1 J from a body at 4.2 K to be rejected to a heat sink at 300 K, when the efficiency of the cooling plant is taken into account.

So using liquid helium means keeping down heat leaks as much as possible: usually liquid nitrogen is used to cool radiation shields surrounding the liquid helium, and the cryostatic apparatus is bulky and costly – so much so that superconducting magnets are the only large-scale application of superconductivity. Though some bubble chamber and accelerator magnets are enormous, there are not many of them and the tonnage of superconducting material used world-wide is relatively small. The econo-

264

mic picture will be vastly changed, however, when superconducting materials capable of carrying reasonable current densities are able to operate at 77 K. In such a case large generators, motors, transformers and transmission lines will be a practical, as opposed to a practicable, proposition, hence the *éclat* with which papers reporting 'high' temperature superconductivity were received in 1986. Table 10.1 gives some data on superconductors and shows the great leap in transition temperatures achieved then.

Table 10.1 Properties of Selected Superconductors in Historical Order

Year	T_c	Material	Class	Crystal Structure	Type	$H_c^\dagger$
1911	4.2	Hg	Metal	Tetragonal	I	0.033
1913	6.2	Pb	Metal	fcc	I	0.064
1930	9.25	Nb	Metal	bcc	II	0.164
1940	15	NbN	I.S.C.[a]	NaCl	II	12.2
1950	17	V_3Si	I.M.C.[b]	W_3O^c	II	12.4
1954	18	Nb_3Sn	I.M.C.[b]	W_3O^c	II	18.5
1960	10	Nb–Ti[d]	Alloy	bcc	II	11.9
1964	0.7	$SrTiO_3$	Ceramic	Perovskite	II	Small
1970	20.7	Nb_3 (Al,Ge)	I.M.C.[b]	W_3O^c	II	34.0
1977	23	Nb_3Ge	I.M.C.[b]	W_3O^c	II	29.6
1986	34	$La_{1.85}Ba_{0.15}CuO_4$	Ceramic	Tetragonal[e]	II	43
1987	94	$YBa_2Cu_3O_{6.8}$	Ceramic	Orthorhombic[f]	II	111[g]

* Superconducting transition temperature in K.
† In MA/m, extrapolated to 0 K. H_{c2} for type II materials.
[a] Interstitial compound.
[b] Intermetallic compound.
[c] The so-called β-tungsten structure.
[d] Commercial magnet wire.
[e] May be slightly orthorhombic.
[f] Nearly tetragonal: $a/b = 1.016$.
[g] Highly anisotropic.

10.2 The Phenomenology of Superconductivity

Superconductivity proved to be fairly widespread among the elements, but the highest transition temperature was only 11 K, in technetium. Commercial exploitation had to wait for the development of stabilized Nb–Ti wire in the 1950s. Onnes's discovery long caused great bafflement to the theoreticians, and the acceptance of the basic explanation had to await the publication of the BCS theory in 1957 (J. Bardeen, L. N. Cooper and J. R. Schrieffer, *Phys. Rev.*, **106**, 162 (1957) and **108**, 1175 (1957)), though its essential characteristic – paired electrons – had been proposed by the chemist R. A. Ogg about a decade before. In the meantime, Meissner and Ochsenfeld discovered in 1933 that a superconductor spontaneously expelled some or all of its magnetic flux as it was cooled below T_c in a magnetic field (the *Meissner effect*). This was quite unexpected: for if $J = \sigma\mathscr{E}$, by Ohm's law, and $\sigma = \infty$ while J is finite, then $\mathscr{E} = 0$; but $\partial B/\partial t \propto \nabla \times \mathscr{E}$, so $\partial B/\partial t = 0$ and the magnetic flux through an infinitely conductive material should not change as it is cooled through T_c. Infinite conductivity was thus insufficient to characterize a superconductor – it must exhibit the Meissner effect too. With the often-impure samples of oxide superconductors now being investigated, the Meissner effect is used as a litmus test for superconductivity.

Figure 10.1 shows the difference between an infinitely conductive material and a superconductor. Both materials will resist flux penetration if a magnetic field is applied below T_c since eddy currents will form and screen out the field. But only a superconductor can expel the flux as it is cooled through T_c. In figure 10.2(a) the plot of magnetization ($M = J/\mu_0$, where J is the magnetic polarization here) against applied field is shown for a long cylindrical sample of pure lead at 4.2 K. The magnetization is equal

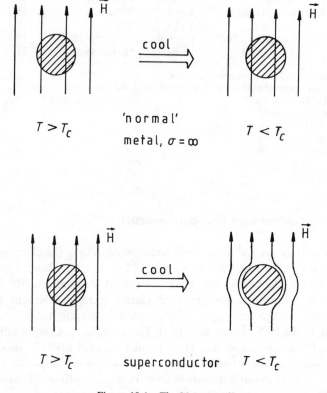

Figure 10.1   *The Meissner effect*

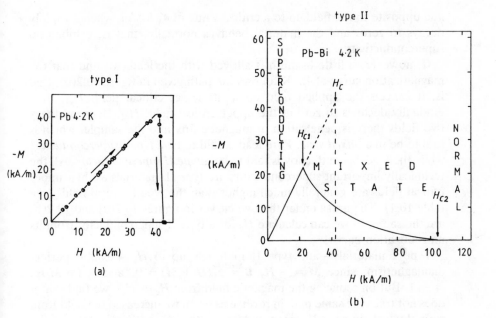

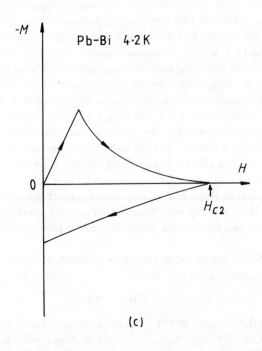

Figure 10.2 *(a) Magnetization curve for pure lead at 4.2 K; (b) magnetization curve for a lead–bismuth alloy at 4.2 K; (c) hysteresis in a type II superconductor*

and opposite to the field up to a critical value of 42 kA/m, when it rapidly decays to zero, and the material behaves normally, that is, exhibits no superconductivity.

If, however, a little bismuth is alloyed with the lead, we find that the magnetization curve at 4.2 K follows the path given in figure 10.2(b), that is, it cancels the applied field up to its lower critical field, H_{c1}, then gradually declines to zero at the upper critical field, H_{c2}. Between these two fields there is penetration of magnetic flux into the sample, which is said to be in a *mixed state*. Pure lead is said to be a *type I superconductor*, while the lead–bismuth alloy is said to be a *type II superconductor*. All the technically important superconductors are type II materials as their upper critical fields are generally much higher than the type I critical fields (see table 10.1). The areas under the two curves in figures 10.2(a) and 10.2(b) are the same, so we can calculate H_c for a type II superconductor from its magnetization curve.

Type I materials – and type II materials up to H_{c1} – exhibit perfect diamagnetism, since $M = -H$, $B = \mu_0(H + M) = 0$, and χ ($\equiv M/H$) $= -1$. But on reducing the magnetic field from H_c or H_{c2}, we find that it does not trace the same path in reverse as when we increased the field from zero. In fact, in type II superconductors, the magnetization follows the path shown in figure 10.2(c) below the field axis, while in type I materials it follows a path a little below the perfectly diamagnetic plot. Thus both samples show hysteresis, but it is much more pronounced for the type II material. Indeed, if a type I superconductor is made from very pure material, and is well annealed so that it contains very few dislocations, then the hysteresis is slight. The hysteresis indicates that the flux which penetrated the sample while the field was increasing is not all expelled when the field is reduced. Cold working a type I material increases the hysteresis considerably, so that dislocations and other defects must cause flux pinning: this is a clue to making hard superconductors that will keep their magnetic flux in place – make them 'dirty'. Movement of magnetic flux (known as *flux-jumping*) is not desirable in a superconducting wire because it causes heating and reversion of superconducting regions to normal conduction, leading sometimes to a runaway condition known as a *quench* in which the whole of a solenoid may revert suddenly to normal conduction.

The critical fields are found to be highly dependent on the temperature, and in type I materials we can write

$$H_c(T) = H_c(0)[1 - (T/T_c)^2] \tag{10.1}$$

where $H_c(T)$ is the critical field at T K and T_c is the transition temperature. For lead, $H_c(4.2) = 42$ kA/m, $T_c = 7.2$ K, so $H_c(0) = 64$ kA/m as given in table 10.1. This relationship is approximately true for the upper critical field of type II materials.

10.3   Characteristic Lengths

In 1935, F. and H. London suggested that Maxwell's equations be supplemented in the case of superconductors by the London equation, which leads to

$$\nabla \times J = - H/\lambda_L^2 \tag{10.2}$$

where λ_L is the *London penetration depth*. Now $\nabla \times H = J$ by Maxwell's equations, where J is the current density, so $\nabla \times \nabla \times H = \nabla \times J$, which is $- H/\lambda_L^2$, from (10.2). It can be shown that $\nabla \times \nabla \times H = - \nabla^2 H$, so that $\nabla^2 H = H/\lambda_L^2$. Considering a superconductor with the field parallel to its surface, whose normal is the x-axis, we can see that

$$H(x) = H(0)\exp(- x/\lambda_L)$$

is a solution if $x = 0$ is the surface and $H(0)$ is the field at the surface. The applied field thus penetrates into the superconductor with exponential decay and characteristic length λ_L. For thin films with thickness, $t << \lambda_L$, the field penetrates uniformly, there is no Meissner effect and the critical field becomes very large. For lead, $\lambda_L = 37$ nm, so the films have to be very thin for H_c to be increased much.

Another characteristic length, which arose from the Ginsberg–Landau theory of about 1950, is known as the *intrinsic coherence length*, ξ_0. Roughly speaking, ξ_0 is the distance over which the superconducting electron concentration remains constant in a varying applied magnetic field. The BCS theory showed that $\xi_0 = \hbar v_f/\pi\Delta$, where v_f is the speed of electrons at the fermi surface and Δ is a difference in superconducting energy states to be discussed later. We can calculate ξ_0 as 83 nm at 0 K for lead (see problem 10.5). In type II superconductors, the coherence length depends on the mean free path of normal conduction electrons, l_m, as do the penetration depth, λ, and ξ_0. Within a factor of $\sqrt{2}$:

$$\xi \approx \sqrt{(\xi_0 l_m)} \qquad \text{and} \qquad \lambda \approx \lambda_L \sqrt{(\xi_0/l_m)}$$

Thus $\lambda/\xi = \kappa = \lambda_L/l_m$, and it is found from Ginsberg–Landau theory that $H_{c1} = H_c/(\kappa\sqrt{2})$ and $H_{c2} = (\kappa\sqrt{2})H_c$, so that $H_{c1}H_{c2} = H_c^2$. When $\kappa < 1$, the superconductor is type I and if $\kappa > 1$ it is type II.

10.3.1   Critical Fields

In a type II superconductor, the magnetic flux penetrates the material so that there are alternate regions of normal and superconducting material as in figure 10.3. In the normal region the field strength is H_a, the same as the applied field, and in the superconducting layer the field strength falls exponentially. The flux in the normal material is quantized in multiples of ϕ_0, called a *fluxoid*, which is given by $\phi_0 = h/2e = 2.07 \times 10^{-15}$ Tm^2. At

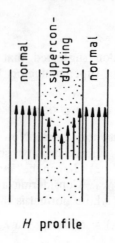

H profile

Figure 10.3 *The field profile in the mixed state*

the lower critical field, H_{c1}, the flux from a fluxoid penetrates a distance λ into the surrounding superconducting material, so the area normal to the flux is roughly $\pi\lambda^2$, and the flux is $\mu_0\pi\lambda^2 H_{c1}$ in T m², which must be equal to ϕ_0, so

$$\mu_0\pi\lambda^2 H_{c1} = \phi_0 = h/2e$$

giving

$$H_{c1} = h/2\pi e\mu_0\lambda^2 = \hbar/e\mu_0\lambda^2$$

Suppose $H_{c1} = 10$ kA/m, then we find $\lambda = 230$ nm. At the upper critical field limit, the fluxoids are packed together as closely as possible and ξ is the smallest possible depth of penetration of the flux into the surrounding superconducting material, giving

$$\pi\xi^2 H_{c2} = \phi_0 = h/2e$$

leading to

$$H_{c2} = \hbar/e\mu_0\xi^2$$

Given that $H_{c2} = 1$ MA/m, we find $\xi = 23$ nm, $\kappa = 10$ and $H_c = 100$ kA/m. These are estimates for the critical fields and differ somewhat from exact calculations based on Ginsberg–Landau theory.

10.4 BCS Theory

In 1950, H. Fröhlich suggested that superconduction might occur through an electron–lattice interaction. Normally electrons are scattered by lattice vibrations giving rise to electrical resistance, but in BCS theory an electron of wavevector k causes a slight distortion in the neighbouring lattice which forms an attractive potential for an electron of wavevector $-k$. In quantum-mechanical terms, the first electron creates a *virtual phonon* and loses momentum, but the second electron then comes along and in colliding with the virtual phonon acquires all the momentum lost by the first electron. The overall momentum change is zero and the paired electrons are superconducting. These *Cooper pairs* are bound together by a very small energy, Δ, forming a new ground state which is superconducting and is separated by an energy gap, 2Δ, from the next lowest (excited) state above it. The fermi level is in the middle of the gap. Δ can be found by measuring the specific heat in both superconducting and normal states (the material is normal below T_c when $H > H_c$) and making an Arrhenius plot of the difference (that is, a plot of $\ln\Delta C_v$ against $1/T$). The slope of the plot is $-\Delta/k_B$. Δ is found to be about $2k_B T_c$ at 0 K, so for niobium $\Delta \approx 1\frac{1}{2}$ meV. Though this energy is very small, so that Cooper pairs are continually being split up, many other electrons are available to form fresh ones, and superconductivity is maintained. BCS theory predicts that the Δ–T graph follows

$$\delta = \tanh(\delta/t)$$

where $\delta = \Delta(T)/\Delta(0)$, $t = T/T_c$ and $\Delta(T)$ is the gap at T K. Thus Δ disappears at T_c and is roughly $\Delta(0)$ for $T < \frac{1}{2}T_c$. The experimental data are in complete accord with BCS theory.

BCS theory is too complicated for discussion here, but it explains all the phenomena described above and makes quantitative predictions which are borne out by experiment; for example, it predicts that the superconducting transition temperature will be

$$T_c = 1.14\Theta\exp[-1/UD(E_f)]$$

where Θ is the Debye temperature, U is the electron–lattice interaction energy and $D(E_f)$ is the electronic density of states at the fermi level. U is higher when the electrical resistivity at 300 K is high, so that T_c is higher for the more resistive elements among those of similar electronic structures and Debye temperatures.

10.5 The Josephson Effect

Brian Josephson was a graduate student when he predicted that when two superconductors were separated by a very thin insulator (1–2 nm), Cooper pairs could tunnel through the insulator and produce a supercurrent flow without any voltage across the junction. Giaever had already shown that 'normal' electrons could cross an insulating barrier under the influence of an applied voltage, the effects predicted by Josephson were all due to supercurrents. These were

(1) The DC Josephson effect: current flow at zero volts.
(2) The AC Josephson effect: a direct voltage causes current oscillations and microwave radiation of frequency, $f = 2Ve/h$. This effect has been used to measure e/h very precisely and has been suggested as a means of establishing a voltage standard.
(3) Quantum interference. If two Josephson junctions are arranged in a ring as in figure 10.4, the current flowing is given by

$$I = I_0 \mid \cos(\pi\phi/\phi_0) \mid$$

where I_0 is a constant and ϕ $(= BA)$ is the flux through the ring. The current varies with ϕ and is a maximum when $\phi = n\phi_0$, $n = 1, 2, 3 \ldots$.

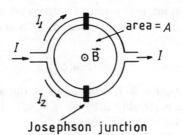

Figure 10.4 *A SQUID*

10.6   High-temperature Ceramic Superconductors

The discovery of superconductivity in $SrTiO_3$ in 1964 opened up a whole new field for superconductivity, since the compound was non-metallic, though its transition temperature was very low. In 1973, a different titanate was found with $T_c = 13$ K, and later other oxide superconductors – none with much higher transition temperatures – were found. More work was done on oxides as a result until a transition temperature of 30 K or more was achieved (J. G. Bednorz and K. A. Mueller, *Z. Phys*, **B64**, 189 (1986)) in an odd compound: $La_{1.85}Ba_{0.15}CuO_4$. Despite colossal efforts by materials scientists, no superconductor with a transition temperature much over 20 K had been found for 16 years, so the impact of the discovery was profound, and set off an explosion in research into oxide superconductivity. Progress was rapid and a transition temperature above the boiling point of nitrogen was soon announced (M. K. Wu *et al.*, *Phys. Rev. Lett.*, **58**, 908 (1987)), in an even odder material, $YBa_2Cu_3O_{7-x}$, where $x \approx 0.2$.

Figure 10.5 shows the structure of $YBa_2Cu_3O_7$, which is *not* a perovskite, though popularly so termed. The unit cell contains one formula unit. Relating it to the perovskite structure, the Ba^{2+} ions occupy O^{2-} sites in a close-packed plane and the copper ions occupy the Ti^{4+} sites of a perovskite, but half the O^{2-} ions in the copper planes are missing. All of the O^{2-} ions are missing in the Y^{3+} planes, causing distortion in the planes above and below (not shown in figure 10.5). The superconductivity is associated with Cu^{2+}/Cu^{3+} charge transfer. Such a complex structure presents a great challenge to theorists and experimentalists.

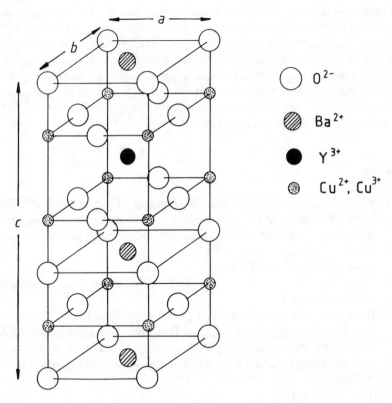

Figure 10.5 *The unit cell of $YBa_2Cu_3O_7$*

10.7 Applications of Superconductivity

The first applications for superconductivity were in solenoidal magnets, though these were a long time coming, because while a short piece of wire might be well-behaved, long pieces were maddeningly unpredictable because of flux-jumping, leading to quenches. By incorporating superconducting wires or layers in a copper (which is normal at 4.2 K) matrix, the reversion of a small part of the wire to normalcy (in which the resistivity is much more than copper's) causes a temporary shunting of current into the copper with no great rise in resistance. The normal material then has time to cool down and become superconducting again. Attempts to exploit superconductivity in motors, generators, transformers and transmission lines all failed the economic test, largely because liquid helium was required. Thus new, high T_c, ceramic materials have the potential to change this, but it will require a great effort to turn them into tractable forms.

There have been a number of other devices based on superconducting quantum effects: Josephson junctions have been used in demonstration high-speed computers, in various electronic circuits and in magnetometers. The SQUID (Superconducting Quantum Interference Device) magnetometer is sold commercially and is in great demand for testing the new ceramic materials. SQUIDs can detect changes in flux down to a fraction of ϕ_0, so a device 3 mm in diameter can detect magnetic fields of $< 10^{-4}$ A/m (the earth's field is about 50 A/m).

Problems

10.1. What is the area under the $M–H$ graph of figure 10.2(a)? This represents the stabilization energy of the superconducting state relative to the normal state of lead, when divided by μ_0. What would be the stabilization energy for lead at 0 K?
[*Ans. 1120 J/m³. 2570 J/m³*]

10.2. Show that $\nabla \times \nabla \times H = - \nabla^2 H$. [See chapter 2, remember $\partial H_x/\partial x = 0$ etc.]

10.3. If the critical current density, J_c, for a superconducting, cylindrical wire is determined by the critical field, and this field is due to J_c, write down an expression for J_c in terms of H_c and wire diameter. If the critical current in a 0.1 mm diameter wire is 100 A, what is the critical field?
[*Ans. 1.27 MA/m*]

10.4. Estimate the lattice parameters for $YBa_2Cu_3O_7$ and hence its density. What is the ratio of Cu^{3+} to Cu^{2+} ions?
[Experimental values: $a = 3.823$ Å, $b = 3.8864$ Å, $c = 11.68$ Å.]

10.5. Calculate ξ_0 for lead. Assuming there are two conduction electrons per atom, estimate Δ and hence T_c.
[*Ans. 83 nm, 1.3 meV, 7.5 K*]

10.6. Show that, because of the demagnetizing field, the magnetization in a superconducting sphere is $-\frac{2}{3}M = H_a$, where H_a is the applied field. Hence show that the field at the surface in the equatorial plane normal to H_a is $1\frac{1}{2}H_a$. [Thus the critical field is reduced to $\frac{2}{3}H_c$ by the demagnetizing field.]

10.7. For a sample of YBa$_2$Cu$_3$O$_{7-x}$, the coherence length, ξ, at 0 K is 1.4 nm and the penetration depth, λ, is 200 nm. What will be H_{c1}, H_{c2} and H_c at 0 K? If $T_c = 94$ K, estimate Δ. Taking $\xi = \xi_0$, what will be the fermi velocity at 0 K? Hence find the carrier concentration at 0 K.

[*Ans. 13.1 kA/m, 268 MA/m, 1.87 MA/m; 16 meV; 107 km/s; 2.67 × 10^{25}/m^3]*

Further Reading

General

C. Kittel, *Introduction to Solid State Physics*, 6th edn, Wiley (1986).
A classic. Advanced undergraduate level.
L. Solymar and D. Walsh, *Lectures on the Electrical Properties of Materials*, 3rd edn, Oxford (1984).

Chapter 1

J. C. Anderson, K. D. Leaver, J. M. Alexander and R. D. Rawlings, *Materials Science*, 2nd edn, Nelson (1974).
J. E. Gordon, *The New Science of Strong Materials*, 2nd edn, Penguin (1976).
Most enjoyable. A popular classic.
H. D. Megaw, *Crystal Structures: A Working Approach*, Saunders (1973).
A clear exposition by a famous female crystallographer.
F. R. N. Nabarro, *The Theory of Crystal Dislocations*, Oxford (1967).
A seminal work by a leading dislocationist.

Chapter 2

J. S. Dugdale, *The Electrical Properties of Metals and Alloys*, Arnold (1977).
G. T. Meaden, *The Electrical Resistance of Metals*, Plenum (1965).
N. F. Mott and H. Jones, *The Theory of the Properties of Metals and Alloys*, Oxford (1936) (reprinted by Dover).
A classic by a Nobel laureate in physics.

276

Chapter 3

H. M. Rosenberg, *The Solid State*, Oxford (1975).
W. Shockley, *Electrons and Holes in Semiconductors*, Van Nostrand (1950).
  Another classic by a Nobel laureate.

Chapter 4

A. Bar-Lev, *Semiconductors and Electronic Devices*, 2nd edn, Prentice-Hall (1984).
D. A. Fraser, *The Physics of Semiconductor Devices*, 4th edn, Oxford (1986).
N. B. Hannay (editor), *Semiconductors*, Reinhold (1959).
  A still-useful work by the materials team of Bell Labs.
S. M. Sze, *Semiconductor Devices*, Wiley (1985).
  A practical work by a contemporary scientist at Bell Labs.

Chapter 5

R. A. Colclaser, *Microelectronics: Processing and Device Design*, Wiley (1980).
N. G. Einspruch (editor), *VLSI Electronics: Microstructure Science*, Academic (1985).
W. S. Ruska, *Microelectronic Processing*, McGraw-Hill (1987).
S. M. Sze (editor), *VLSI Technology*, 2nd edn, McGraw-Hill (1989).
Annual Book of ASTM Standards, part 43 Electronics, ASTM.

Chapter 6

R. M. Bozorth, *Ferromagnetism*, Van Nostrand (1951).
  A classic. Contains all there is to know about magnetic materials up to 1950.
J. P. Jakubovics, *Magnetism and Magnetic Materials*, Institute of Metals (1987).
E. P. Wohlfarth (editor), *Ferromagnetic Materials*, vols 1–4, North-Holland (1978–1988).
  The successor to Bozorth. Wohlfarth died just before completing this major work.

Chapter 7

H. Jouve (editor), _Magnetic Bubbles_, Academic (1986).

M. McCaig, _Permanent Magnets in Theory and Practice_, Pentech (1977). A legacy of the Permanent Magnet Association, which disbanded in 1974.

J. C. Mallinson, _The Foundations of Magnetic Recording_, Academic (1987).

S. Smith, _Magnetic Components: Design and Applications_, Van Nostrand-Reinhold (1984).

E. C. Snelling, _Soft Ferrites: Properties and Applications_, 2nd edn, Butterworths (1988).

Chapter 8

H. Frölich, _The Theory of Dielectrics_, Oxford (1958).

P. J. Harrop, _Dielectrics_, Butterworths (1972).

B. Jaffe, W. R. Cook and H. Jaffe, _Piezoelectric Ceramics_, Academic (1971). Mainly on perovskites.

Chapter 9

W. B. Jones, _Introduction to Optical Fiber Communication Systems_, Holt, Rinehart and Winston (1988).

J. Wilson and J. F. B. Hawkes, _Optoelectronics: an Introduction_, 2nd edn, Prentice-Hall (1989). Covers almost everything in the field.

A. Yariv, _Optical Electronics_, 3rd edn, Holt, Rinehart and Winston (1985).

Chapter 10

A. C. Rose-Innes and E. H. Rhoderick, _Introduction to Superconductivity_, 2nd edn, Pergamon (1978).

C. W. Turner, _The Principles of Superconductive Devices and Circuits_, Arnold (1981).

A. Barone and G. Paterno, _The Physics and Applications of the Josephson Effect_, Wiley (1982).

S. A. Wolf and V. Z. Kresin (editors), _Novel Superconductivity_, Plenum (1987).

Appendix: The Periodic Table of the Elements

IA	IIA	IIIB	IVB	VB	VIB	VIIB	VIII	VIII

Legend (key box):

Chemical symbol	**Be** 4	Atomic No.
Atomic weight	9.01	
Crystal structure	A2 2.55	Lattice parameter (Å)
(upper is higher	A3 2.29	(a, c for A3)
temp allotrope)	3.58	
Outer electrons	1s²2s²	

Structures:
- A1 fcc
- A2 bcc
- A3 hcp
- A4 diamond
- A5 bct
- A6 fct
- A7 arsenic
- A8 selenium

- C cubic
- H hexagonal
- M monoclinic
- O orthorhombic
- R rhombohedral
- T tetragonal

Period 1–3 (Groups IA, IIA):

IA	IIA
H 1 1.008 A3 3.76 6.13 1s¹	
Li 3 6.94 A2 3.51 A3 3.11 5.09 1s²2s¹	**Be** 4 9.01 A2 2.55 A3 2.29 3.58 1s²2s²
Na 11 22.99 A2 4.29 2p⁶3s¹	**Mg** 12 24.31 A3 3.21 5.21 2p⁶3s²

Period 4:

K 19	**Ca** 20	**Sc** 21	**Ti** 22	**V** 23	**Cr** 24	**Mn** 25	**Fe** 26	**Co** 27
39.10	40.08	44.96	47.90	50.94	52.00	54.94	55.85	58.93
A2 5.25	A2 4.48	A3 3.31	A2 3.31	A2 3.03	A1 3.68	A1 3.86	A2 2.93	A1 3.54
	A1 5.58	5.27	A3 2.95		A2 2.89	C 6.32	A1 3.65	A3 2.51
			4.68			C 8.92	A2 2.97	4.07
3p⁶4s¹	3p⁶4s²	3d¹4s²	3d²4s²	3d³4s²	3d⁴4s²	3d⁵4s²	3d⁶4s²	3d⁷4s²

Period 5:

Rb 37	**Sr** 38	**Y** 39	**Zr** 40	**Nb** 41	**Mo** 42	**Tc** 43	**Ru** 44	**Rh** 45
85.47	87.62	88.91	91.22	92.91	95.94	97	101.07	102.91
A2 5.61	A3 4.32	A2 4.11	A2 3.61	A2 3.30	A2 3.15	A3 2.74	A3 2.71	A1 3.80
	7.06	A3 3.65	A3 3.23			4.40	4.28	
	A1 6.09	5.73	5.15					
4p⁶5s¹	4p⁶5s²	4d¹5s²	4d²5s²	4d⁴5s¹	4d⁵5s¹	4d⁵5s²	4d⁷5s¹	4d⁸5s¹

Period 6:

Cs 55	**Ba** 56	**La†** 57	**Hf** 72	**Ta** 73	**W** 74	**Re** 75	**Os** 76	**Ir** 77
132.91	137.34	138.91	178.49	180.95	183.85	186.2	190.2	192.2
A2 6.14	A2 5.02	A2 4.26	A2 3.50	A2 3.30	A2 3.17	A3 2.76	A3 2.74	A1 3.84
		A1 5.30	A3 3.20			4.46	4.32	
		H 3.77	5.05					
		5.05						
5p⁶6s¹	5p⁶6s²	5d¹6s²	5d²6s²	5d³6s²	5d⁴6s²	5d⁵6s²	5d⁶6s²	5d⁷6s²

Period 7:

Fr 87	**Ra** 88	**Act‡** 89
223	226	227
		A1 5.31
6p⁶7s¹	6p⁶7s²	6d²7s²

† Rare Earths:

La 57	**Ce** 58	**Pr** 59	**Nd** 60	**Pm** 61	**Sm** 62	**Eu** 63
138.9	140.1	140.9	144.2	147	150.4	152.0
A2 4.26	A1 5.16	A2 4.12	A2 4.13		A2 4.07	A2 4.58
H 3.77	H 3.68	H 3.67	H 3.66		R 8.98	
12.17	11.92	11.04	11.80			
5d¹6s²	4f²6s²	4f³6s²	4f⁴6s²	4f⁵6s²	4f⁶6s²	4f⁷6s²

‡ Actinides:

Ac 89	**Th** 90	**Pa** 91	**U** 92	**Np** 93	**Pu** 94	**Am** 95
227	232	231	238	237	244	243
A1 5.31	A2 4.11	T 3.93	A2 3.53	A2 3.52	O 3.16	H 3.64
	A1 5.08		T 10.8	T 4.90	M 9.28	11.76
			O 2.85	O 4.72	M 6.18	
6d²7s²	6d³7s²	5f²6d¹7s²	5f³6d¹7s²	5f⁵7s²	5f⁶7s²	5f⁷7s²

VIII	IB	IIB	IIIA	IVA	VA	VIA	VIIA	0
								He 2 4.003 A3 3.58 5.84 A2 1s²

			B 5
			10.61
			H
			T
			R 9.45
			2s²2p¹
				C 6
				12.01
				A4 3.57
				H 2.46
				      6.71
				2s²2p²
					N 7
					14.01
					H 4.04
					      6.63
					C 5.66
					2s²2p³
						O 8
						16.00
						C 6.83
						R 4.21
						
						2s²2p⁴
							F 9
							19.00
							
							2s²2p⁵
								Ne 10
								20.18
								A1 4.46
								
								2s²2p⁶

			Al 13 26.98 A1 4.05 3s²3p¹	**Si 14** 28.09 A4 5.43 3s²3p²	**P 15** 30.97 C 11.31 O 3.32 C 18.8 3s²3p³	**S 16** 32.06 O 10.41 M 10.92 R 6.45 3s²3p⁴	**Cl 17** 35.45 A1 A3 A2 3s²3p⁵	**Ar 18** 39.95 A1 5.47 A1 5.31 3s²3p⁶
Ni 28 58.71 A1 3.52 3d⁸4s²	**Cu 29** 63.54 A1 3.62 3d¹⁰4s¹	**Zn 30** 65.37 A3 2.67 4.95 3d¹⁰4s²	**Ga 31** 69.72 O 4.53 4s² 4p¹	**Ge 32** 72.59 A4 5.66 4s²4p²	**As 33** 74.92 A7 4.13 4s²4p³	**Se 34** 78.96 A8 4.37 M 9.05 M 12.85 4s²4p⁴	**Br 35** 79.91 O 4.48 4s²4p⁵	**Kr 36** 83.80 A1 5.64 4s²4p⁶

Pd   46
106.4
A1   3.89
 	
4p⁶4d¹⁰
	Ag 47
	107.9
	A1 4.09
	
	4d¹⁰5s¹
		Cd 48
		112.4
		A3 2.67
		      4.95
		
		4d¹⁰5s²
			In 49
			114.8
			A6 4.60
			      4.95
			
			5s²5p¹
				Sn 50
				118.7
				A5 5.83
				      3.18
				A4 6.49
				5s²5p²
					Sb 51
					121.8
					A7 4.51
					      57.1
					
					5s²5p³
						Te 52
						127.6
						A8 4.46
						
						5s²5p⁴
							I 53
							126.9
							O 4.79
							
							5s²5p⁵
								Xe 54
								131.3
								A1 6.13
								
								5s²5p⁶

Pt 78 195.1 A1 3.92 5d⁹6s¹	**Au 79** 197.0 A1 4.08 5d¹⁰6s¹	**Hg 80** 200.6 R 3.01 T 4.00 5d¹⁰6s²	**Tl 81** 204.4 A2 3.88 A3 3.46 5.53 6s²6p¹	**Pb 82** 207.2 A1 4.95 6s²6p²	**Bi 83** 209.0 A7 4.54 6s²6p³	**Po 84** 209 R 3.36 C 3.35 6s²6p⁴	**At 85** 210 6s²6p⁵	**Rn 86** 222 6s²6p⁶

Gd 64	Tb 65	Dy 66	Ho 67	Er 68	Tm 69	Yb 70	Lu 71

157.3
A3   3.63
      5.78
4f⁷5d¹6s²
	158.9
	A3 3.61
	      5.70
	4f⁸5d¹6s²
		162.5
		A3 3.59
		      5.65
		4f¹⁰6s²
			164.9
			A3 3.58
			      5.62
			4f¹¹6s²
				167.3
				A3 3.56
				      5.59
				4f¹²6s²
					168.9
					A3 3.54
					      5.55
					4f¹³6s²
						173.0
						A2 4.44
						H 5.48
						4f¹⁴6s²
							175.0
							A3 3.51
							      5.55
							4f¹⁴5d¹6s²

Cm 96	Bk 97	Cf 98	Es 99	Fm 100	Md 101	No 102	Lr 103
247 5f⁷6d¹7s²	247 5f⁸6d¹7s²	251 5f¹⁰7s²	254 5f¹¹7s²	257 5f¹²7s²	256 5f¹³7s²	254 5f¹⁴7s²	257 6d¹7s²

Index

282